ANIMAL AND PLANT
Anatomy

VOLUME CONSULTANTS

• Amy-Jane Beer, *Natural history writer and consultant* • Valerius Geist, *University of Calgary, Alberta, Canada*

• John Gittleman, *University of Virginia, VA* • Tom Jenner, *Academia Británica Cuscatleca, El Salvador*

• Ray Perrins, *Bristol University, England* • Adrian Seymour, *Bristol University, England*

• David Spooner, *University of Wisconsin, Madison, WI*

7

Penguin – Reproductive system

 **Marshall Cavendish**
Reference
New York

CONTRIBUTORS

Roger Avery; Richard Beatty; Amy-Jane Beer; Erica Bower; Trevor Day; Erin Dolan; Bridget Giles; Natalie Goldstein; Tim Harris; Christer Hogstrand; Rob Houston; John Jackson; Tom Jackson; James Martin; Chris Mattison; Katie Parsons; Ray Perrins; Kieran Pitts; Adrian Seymour; Steven Swaby; John Woodward.

CONSULTANTS

Barbara Abraham, Hampton University, VA; Glen Alm, University of Guelph, Ontario, Canada; Roger Avery, Bristol University, England; Amy-Jane Beer, University of London, England; Deborah Bodolus, East Stroudsburg University, PA; Allan Bornstein, Southeast Missouri State University, MO; Erica Bower, University of London, England; John Cline, University of Guelph, Ontario, Canada; Trevor Day, University of Bath, England; John Friel, Cornell University, NY; Valerius Geist, University of Calgary, Alberta, Canada; John Gittleman, University of Virginia, VA; Tom Jenner, Academia Británica Cuscatleca, El Salvador; Bill Kleindl, University of Washington, Seattle, WA; Thomas Kunz, Boston University, MA; Alan Leonard, Florida Institute of Technology, FL; Sally-Anne Mahoney, Bristol University, England; Chris Mattison; Andrew Methven, Eastern Illinois University, IL; Graham Mitchell, King's College, London, England; Richard Mooi, California Academy of Sciences, San Francisco, CA; Ray Perrins, Bristol University, England; Kieran Pitts, Bristol University, England; Adrian Seymour, Bristol University, England; David Spooner, University of Wisconsin, WI; John Stewart, Natural History Museum, London, England; Erik Terdal, Northeastern State University, Broken Arrow, OK; Phil Whitfield, King's College, University of London, England.

Marshall Cavendish
99 White Plains Road
Tarrytown, NY 10591–9001

www.marshallcavendish.us

© 2007 Marshall Cavendish Corporation

Library of Congress Cataloging-in-Publication Data
Animal and plant anatomy.
    p. cm.
ISBN-13: 978-0-7614-7662-7 (set: alk. paper)
ISBN-10: 0-7614-7662-8 (set: alk. paper)
ISBN-13: 978-0-7614-7670-2 (vol. 7)
ISBN-10: 0-7614-7670-9 (vol. 7)
1. Anatomy. 2. Plant anatomy. I. Marshall Cavendish Corporation. II.
Title.

QL805.A55 2006
571.3--dc22

                                        2005053193

Printed in China
09 08 07 06 1 2 3 4 5

MARSHALL CAVENDISH

Editor: Joyce Tavolacci
Editorial Director: Paul Bernabeo
Production Manager: Mike Esposito

THE BROWN REFERENCE GROUP PLC

Project Editor: Tim Harris
Deputy Editor: Paul Thompson
Subeditors: Jolyon Goddard, Amy-Jane Beer, Susan Watts
Designers: Bob Burroughs, Stefan Morris
Picture Researchers: Susy Forbes, Laila Torsun
Indexer: Kay Ollerenshaw
Illustrators: The Art Agency, Mick Loates, Michael Woods
Managing Editor: Bridget Giles

Contents

Penguin

CLASS: Aves ORDER: Sphenisciformes FAMILY: Spheniscidae

Penguins are so unlike other birds that the first European seafarers to encounter them called them "feathered fish." Penguins have lost the power of flight during their evolution and now have a body perfectly suited to an underwater lifestyle in cold seas. Their fat, torpedo-shape form slips easily through the ocean, yet is bulky enough to store plenty of food reserves for the lean times when the birds are confined to land for breeding. The penguin family is at least 50 million years old, but natural selection has conserved the very successful penguin shape and black-and-white color. The 17 species alive today differ mainly in size and head color.

▼ *Penguins are not closely related to any other birds and are placed in a group of their own, the order Sphenisciformes. The six genera of penguins are shown.*

Anatomy and taxonomy

All life-forms are classified in groups of closely related species. Classification is based mainly on shared anatomical and genetic features, which usually indicate that the members of a group have the same ancestry. Thus the classification shows how life-forms are related to each other.

● **Animals** All animals are multicellular and depend on other organisms for food. They differ from other multicellular life-forms in their ability to move around (generally using muscles) and respond rapidly to stimuli.

● **Chordates** At some time in its life cycle a chordate has a stiff, dorsal (back) supporting rod called the notochord. All vertebrates (fish, reptiles, birds, amphibians, and mammals) are chordates, but so are sea squirts and other lesser-known invertebrates. Most chordates have a row of chevron-shaped blocks of muscle called myotomes running down their notochord. This feature is especially obvious in fish.

● **Vertebrates** Chordates whose notochord changes into a backbone during the development of the embryo are vertebrates. They include fish, reptiles, amphibians, birds, and mammals. A backbone, or vertebral column, is made up of a chain of small elements called vertebrae, made of cartilage or bone. Vertebrates have a braincase, or cranium, which gives them their alternative name, craniates.

Animals
KINGDOM Animalia

Chordates
PHYLUM Chordata

Vertebrates
SUBPHYLUM Vertebrata

Birds
CLASS Aves

Penguins
ORDER Sphenisciformes

FAMILY Spheniscidae

African and South American penguins	Crested penguins	Blue penguins	Large antarctic penguins	Yellow-eyed penguin	Adélie, gentoo, and chinstrap penguins
GENUS *Spheniscus*	GENUS *Eudyptes*	GENUS *Eudyptula*	GENUS *Aptenodytes*	GENUS AND SPECIES *Megadyptes antipodes*	GENUS *Pygoscelis*

● **Birds** The class Aves comprises vertebrates with skin covered by feathers, which usually occur in bands called feather tracts. Birds have a hollow, lightweight skeleton suited to powered flight. The forelimb bones have fused or disappeared during evolution to leave a single strong strut supporting a wing of extended feathers. The breastbone is expanded into a sail-like keel, which anchors huge flight muscles that power the wing beat. Modern birds have no teeth and instead have a horny and bony beak. Like mammals, birds are homeothermic ("warm-blooded")—that is, they maintain a constant, high body temperature independent of ambient temperature.

● **Penguins** Penguins are so different from other living birds that they are placed in their own group, the order Sphenisciformes, or penguin-shaped birds. The order contains only one living family, the Spheniscidae, or penguin family. Members of the family possess a unique set of features distinguishing them from other birds. They have a heavy, solid skeleton without air spaces. Penguins are flightless, and their wings form stiff paddles with which they propel themselves through water. Penguins' feathers are short, stiff, interlocking, and waterproof. The feathers are not confined to feather tracts, as they are in other birds, but cover the whole body. Their legs are set so far back on their streamlined, torpedo-shaped body that they walk upright on land.

● **Large Antarctic penguins** The genus *Aptenodytes* contains the two largest living penguins: the king penguin, which reaches a height of 3 feet (0.9 m) from bill to tail, and the 4-feet-tall (1.2 m) emperor penguin. They each have a long, down-curved beak; bright yellow auricular (ear) patches; and purple or lilac mandibular

▲ *The yellow-eyed penguin is one of the rarest species, with a population of about 1,500 breeding pairs. These penguins nest in coastal forest and dive for small fish in the ocean around New Zealand's South Island.*

(jaw) plates. Both species live in the Antarctic. The king penguin breeds on subantarctic islands such as South Georgia, and the emperor breeds only on sea ice around mainland Antarctica. The emperor penguin is the largest penguin and the only bird in the world that does not return to land to breed.

FEATURED SYSTEMS

EXTERNAL ANATOMY Penguins have a streamlined, waterproof body, which is propelled through the ocean by stiff wings acting as flippers. Their stance on land, however, is uniquely upright. *See pages 870–872.*

SKELETAL SYSTEM Relative to other birds, penguins have a heavy, solid skeleton. The wing bones are flattened and fused into a rigid flipper shape. *See pages 873–874.*

MUSCULAR SYSTEM Penguins retain the large flight muscles of flying birds but use them to "fly" underwater. The muscles can store more oxygen for long underwater dives. *See page 875.*

NERVOUS SYSTEM Penguins are underwater predators guided by eyesight, but they can see equally well in air. Their eyes are sensitive to dim light. *See page 876.*

CIRCULATORY AND RESPIRATORY SYSTEMS Without specialized circulation and respiration, penguins could not perform their feats of endurance, diving in cold, polar seas. Like other birds, penguins have accessory air sacs and unidirectional lungs, but they also have heat regulation mechanisms in the wings, feet, and face. *See pages 877–878.*

DIGESTIVE AND EXCRETORY SYSTEMS Penguins' digestive system must cope with long periods of fasting every year, when the birds' molting or breeding prevent them from making foraging trips in the ocean. *See page 879.*

REPRODUCTIVE SYSTEM To complete their life cycle, some penguins have unique features. They build no nest and balance their eggs on their feet, and male emperor penguins produce a kind of "milk." *See pages 880–881.*

External anatomy

CONNECTIONS

COMPARE a penguin's coloring with that of an **ALBATROSS**, which is another counter-shaded seabird.

COMPARE a penguin's swimming limbs with those of a **SEAL**, another ocean animal with land-based ancestors.

COMPARE the penguin's short, stiff feathers with the long flight feathers of an **EAGLE**.

Despite differences in size and head decoration, the 17 species of penguins have a consistent body form. They appear squat on land, but in the ocean they form a smooth, pointed wedge that cuts easily through the water. The first species of penguin to be described scientifically, the African penguin, was named *Pygoscelis*, meaning "little wedge," for its streamlined shape.

Penguins obey Bergmann's rule, which predicts that the nearer to the poles animals live, the larger they will be. The smallest penguins, the Galápagos and blue penguins, live in the warmest climate, farthest from the South Pole; and the largest penguins, the emperors, live nearest the South Pole. Larger animals have a smaller surface area relative to their mass, so they can conserve heat more effectively in a cold climate. Emperors further reduce their surface area by having a relatively short bill and wings. If these were longer they would be particularly vulnerable to freezing.

Underwater fliers

The penguin's body is well suited to its ocean lifestyle. It is an underwater predator of fish, shrimplike krill, and squid, and it spends most of its life at sea, often swimming and diving underwater. When foraging, smaller penguin species such as macaroni penguins routinely make dives to around 160 feet (50 m) lasting about 60 seconds. However, Adélie, king, and emperor penguins are all deep divers. The biggest, the emperor penguin, can dive for 18 minutes and reach a depth of 1,750 feet (534 m). Most other diving birds, such as ducks, grebes, loons, and cormorants, propel themselves with their feet, but penguins use their webbed feet as rudders, for steering only. The power comes from their flapping wings, which are stiff and flipperlike. Penguins use a swimming method that many experts call "underwater flying." Just as the wings act as airfoils when a bird is flying in air, penguin wings act as hydrofoils in water. The shape of an airfoil in cross section is such that when it moves through air, it generates an upthrust called lift. Although lift is not important in the dense medium of water, penguin wings generate thrust in the same way that bird wings do in aerial flight.

A penguin's legs are placed so far back on its body that it has a uniquely upright stance on land, and the legs are so short that it has an awkward gait when walking. Penguins traveling long distances over ice and snow move faster and more efficiently by tobogganing: they lie on their belly and push with their toes. Their smooth, fat belly gives a friction-free tobogganing action, but its main

▶ PORPOISING

Adélie penguin
Penguins are able to travel through water at up to 25 miles per hour (40 km/h) by a combination of swimming and porpoising, or leaping clear of the water. Porpoising allows penguins to breathe and may help them avoid predators.

▼ Emperor penguin

All penguins have a similar body shape, and most have similar coloration. Apart from the blue penguin, which is blue-gray, all penguin species are black or dark gray above, with a white or very pale underside.

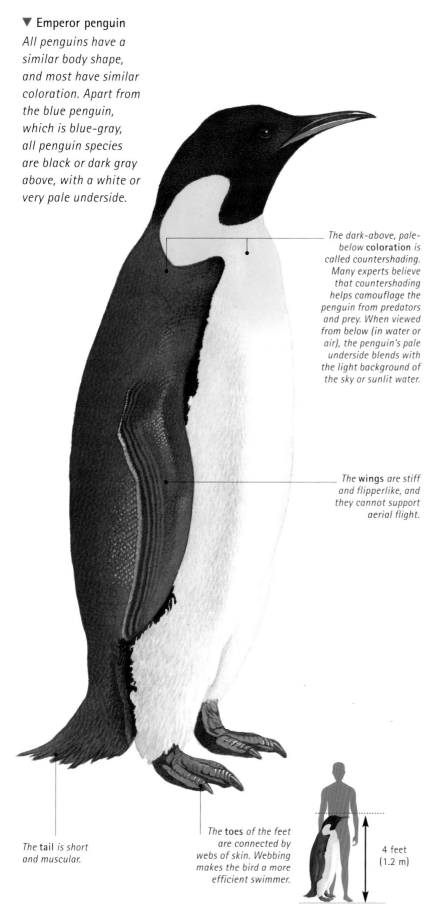

The dark-above, pale-below coloration is called countershading. Many experts believe that countershading helps camouflage the penguin from predators and prey. When viewed from below (in water or air), the penguin's pale underside blends with the light background of the sky or sunlit water.

The wings are stiff and flipperlike, and they cannot support aerial flight.

The tail is short and muscular.

The toes of the feet are connected by webs of skin. Webbing makes the bird a more efficient swimmer.

4 feet (1.2 m)

function is energy storage. Penguins lay down thick reserves of fat under their skin for annual periods of enforced starvation, during which they cannot hunt because they are busy molting their feathers or incubating eggs.

Feather markings

Any distinctive markings on a penguin are usually on its head and neck, perhaps because those are the only parts visible from above when the bird is floating in water. The crested penguins, such as the macaroni, rockhopper, and Fiordland crested penguins, bear their characteristic tufts above each eye, while the African, Magellanic, Humboldt, and Galápagos penguins are distinguished by the specific pattern of black-and-white banding around their head and neck.

What unites penguins, though, is their thick, dense covering of feathers. Since penguins are warm-blooded animals, maintaining a warm body temperature in some of the coldest places on Earth, they need a lot of insulation to keep them from losing heat. Whereas the feathers of other birds grow from bands called feather tracts, or pteryloses, a penguin's feathers sprout

CLOSE-UP

Feathers and molting

Penguin feathers are unique in structure. There is no division into downy feathers and flight feathers as in other birds. All the feathers are small, stiff, and lanceolate (lance-shaped). The longest part is the rachis, or shaft. Beneath it is a thick, downy tuft of fibers called a hyporachis, which makes up for the lack of downy feathers. Like all birds, penguins must molt their old feathers and grow new ones. A penguin molts quickly, because it is poorly insulated without a dense covering of new plumage. Its insulation is useless underwater during this time, so a penguin cannot hunt and cannot eat until molting is over. The feathers of most birds fall out before the new ones grow in their place. Not so with penguins: the shafts of the new feathers push the old feathers out as they grow, leaving no waiting period between loss of the old and growth of the new.

871

EVOLUTION

Penguin origins

The fossil history of penguins stretches back to the Eocene period, 45 million years ago. The earliest penguins closely resemble living penguins, so they do not provide any clues on how penguins became flightless. Because penguins "fly" underwater (that is, they swim by flapping their wings), their ancestors probably resembled today's underwater fliers, such as diving petrels and auks. Owing to the differing demands on the wing, efficient aerial flight is compatible with underwater flight only in small birds of less than 2 pounds (1 kg) in weight. The penguins' flying ancestor must therefore have been puffin-size or smaller. As soon as this ancestor lost the ability to fly, however, it was free to grow heavier and more specialized for diving. There was a rapid growth and diversification of penguin species in the Eocene, with up to 40 fossil species so far discovered, some of which stood as tall as an adult human (up to 5.9 feet, 180 cm).

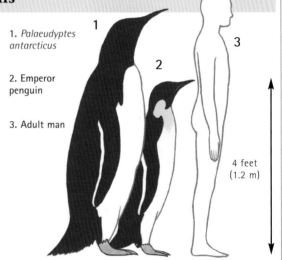

1. *Palaeudyptes antarcticus*

2. Emperor penguin

3. Adult man

4 feet (1.2 m)

from nearly every part of its skin and grow much more densely that those of other birds. The emperor penguin has the highest feather density of any bird, with about 100 feathers per square inch (15.4 per cm²). A penguin must maintain the insulating properties of its plumage by preening; it produces an oily secretion in the preen gland at the base of the tail, wipes its bill on the secretion, and then runs the bill through its feathers.

The feather layer is a much more effective heat insulator than the layer of fat beneath, and provides between 80 and 90 percent of its insulation. The feathers also form a waterproof barrier and trap a layer of warm air against the skin when the penguin is underwater. Water conducts heat away from the penguin's body much more quickly than air, so insulation underwater is doubly important. Dense and overlapping, the feathers provide a barrier resistant to water and extreme cold (to −65°F, or −54°C), and they remain smooth and compact in winds of up to 40 miles per hour (60 km/h).

▶ *An adult emperor penguin guards a group of young birds. Emperor penguin chicks are often adopted temporarily by unrelated adults and gathered into a crèche. The adopting parents are usually birds without a mate or pairs with no young of their own. Typically, a crèche lasts for no more than two days.*

Skeletal system

The skeleton of penguins has many typically birdlike features. In many parts of the body, bones have disappeared or fused during the animal's long evolution. This process has left fewer bones, which make up a strong, reinforced skeleton. The tail of ancient birds such as archaeopteryx was long and composed of many vertebrae. However, a penguin's tail is a stumpy projection of fused vertebrae called a pygostyle. Those penguins that appear to have a long tail, such as the *Pygoscelis* penguins (Adélie, gentoo, and chinstrap penguins), simply have longer tail feathers—their pygostyle is just as stumpy. Farther up the backbone, the thoracic, lumbar, and sacral vertebrae have fused to form a single bone called the synsacrum, which itself is fused to a completely fused pelvic girdle. The fusion of bones has resulted in a strong and inflexible central region in a bird's body. In addition, a bird's skull rides on a chain of 15 neck vertebrae with very mobile joints, allowing great freedom of movement and reach for the head.

Swimming apparatus

The wing bones of birds are also fused. Birds have no separate carpals (hand bones), and their metacarpals and phalanges (hand and finger bones) have fused into a single carpometacarpus bone. This bone acts as a stiff leading edge to the wing. Penguins, which use their wings for swimming, have wings that are not only fused but also flattened into a rigid flipper shape. There is such a reduction in the mobility of the wing joints that a penguin cannot fold its wings against its body. The only really mobile joint is that between the humerus (upper arm) and the shoulder girdle. Rigidity enables the wing to resist being deformed by the dense medium of water, and allows the penguin to propel itself with the upstroke as well as the

downstroke. If the wings were more flexible, they would crumple on the upstroke and be useless for propulsion.

Like most other birds, penguins have an enormous sternum (breastbone) with a sail-like keel projecting forward and downward. Since the keel anchors the massive flight muscles of flying birds, it might seem strange that penguins also have a keeled sternum. However, penguins are not really flightless. Their method of swimming can be regarded as flying underwater. They use an action similar

CONNECTIONS

COMPARE the penguin's V-shape furcula (wishbone) with that of an *EAGLE*. An eagle needs a U-shape wishbone for its soaring flight.

COMPARE the penguin's flat, plantigrade, palmate (forward-pointing toe bones) feet with a *HUMMINGBIRD*'s, which are used for perching. The hummingbird's feet have two backward-pointing toes and two forward-pointing toes, all of which are curved.

COMPARE the penguin's sternum with that of an *OSTRICH*. The ostrich's sternum has no keel because it has no need to anchor flight muscles.

The **bill** is covered with horny, bony plates. The emperor penguin's bill is long and thin—ideal for catching fish.

There are 15 cervical vertebrae.

scapula

femur

ilium

There are 6 caudal vertebrae

pygostyle

furcula (wishbone)

humerus

radius

ulna

The **sternum** is deeply keeled.

The leading edge of the wing is formed by the carpometacarpus.

tibiotarsus

fibula

The feet are palmate, with four forward-pointing toes. The fifth digit is small and does not have a function (it is a vestigial feature).

The **tarsometatarsus** is characteristically short and thick (it is long and thin in other birds).

◀ Emperor penguin
Note how the legs are situated far back on the body, giving the bird an upright stance. The coracoids, or shoulder bones, do not move up and down with the wing. Instead, they remain in place and act as braces. Coracoid bones are absent in mammals.

▲ *Although penguins are flightless, their wings are useful for balance on land and for underwater propulsion. These Adélie penguins are using their wings to balance on an iceberg.*

to, and the same muscles as, birds that fly in air. The arrangement differs a little, however. The furcula (wishbone), for example, is markedly V-shape, rather than U-shape as it is in soaring birds, and it is suspended above rather than fused with the sternum.

Not pneumatic

A penguin's skeleton differs most from that of other birds in its weight. Flying birds have a super-lightweight skeleton achieved by bones filled with a honeycomb of air spaces or by bones that are hollow, thin, and delicate. In some bones, the air spaces are continuous with the air spaces of the respiratory system, so air entering the lungs can pass into the bones. Such a skeleton is called "pneumatic," but penguins have no need for a pneumatic

EVOLUTION

Devotion to diving

A key stage in penguins' evolution was their loss of airborne flight. Birds such as auks and diving petrels, which perform both aerial flight and underwater flight, must compromise their wing design for movement in the two media, air and water. When penguins no longer flew in air, natural selection was free to optimize their wings for underwater flight. Penguin wings no longer had to generate lift, so they did not need to fold on the upstroke. They could remain stiff and provide propulsion when flapping both upward and downward; nor did the wings need to fold for rapid aerobatic flight maneuvers. The density of water is so great that changing body posture is sufficient for maneuvering underwater.

▼ WING BONES
Compare the narrow wing bones of flying seabirds, such as a gull and a razorbill, with the more robust bones of birds that "fly" underwater, such as the extinct great auk, Lucas auk (Mancalla), and a penguin.

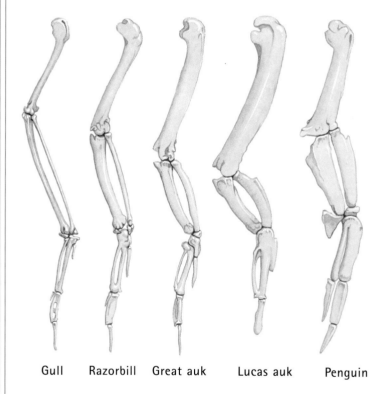

| Gull | Razorbill | Great auk | Lucas auk | Penguin |

skeleton. In fact, an air-filled skeleton would hinder them when diving because it would be too buoyant, and they would have to work hard to overcome their buoyancy. Because they have such heavy bones, penguins sit very low in the water when floating at the surface, and only their head and neck remains above water.

Muscular system

The largest muscles in any flying bird are the massive flight muscles. Penguins' flight muscles are as large as those of flying birds, and the muscles and their tendons are stronger. In emperor penguins, these muscles are four times more massive than the leg muscles. Penguins need large muscles to power their underwater flight. The biggest muscles of all, the pectoralis majors, power the downstroke of the wings, which provides most of the thrust while swimming. Each pectoralis major attaches to the downward-pointing keel of the sternum and extends to the underside of the humerus (upper arm bone), pulling it down as the muscle contracts. For the upstroke, the penguin uses the supracoracoideus. Although it also produces thrust, the upstroke is not as powerful, and the supracoracoideus is not as large. It, too, attaches to the keel of the sternum, but it runs to the dorsal (upper) side of the humerus to pull it up.

Features of penguin muscles allow them to generate heat without shivering—a process called nonshivering thermogenesis. Mammals can generate heat without shivering by breaking down their brown adipose (fatty) tissue, releasing chemical energy. Penguins, like other birds, do not have brown adipose tissue, but their skeletal muscle contains unusual mitochondria. Mitochondria are tiny organelles (mini-organs) within cells that are the site of cell respiration—the process by which the body uses oxygen to release chemical energy from its stores of energy. The muscles in a penguin's body have mitochondria with greatly folded internal membranes, which can drive the heat-generating chemical processes quickly.

Penguins are great divers, but since they are air-breathing animals they cannot take in oxygen while they are underwater. Instead, they must rely on oxygen stores in their air passages, blood, and muscles. Muscles are packed with a protein called myoglobin, which can store oxygen. Penguin myoglobin has a high oxygen-storage capacity, and penguins have a lot of it. Emperor penguins' swimming muscle is 6.4 percent myoglobin by mass,

IN FOCUS

Feather muscles

Penguins have a tiny muscle attached to each feather. It pulls the feather erect when the penguin is on land. The penguin's plumage fluffs up to trap a thicker layer of air for greater insulation. Underwater, the feather muscles relax and are allowed to lie flatter against the penguin's body. A thick layer of air would make the penguin too buoyant underwater.

COMPARE the simple arrangement of the two major pairs of flight muscles in the penguin with the complex arrangement of many flight muscles in the *FRUIT BAT*.

CONNECTIONS

among the highest percentages of any species. Its swimming muscles provide a greater oxygen reserve than the penguins' entire volume of blood. Penguin muscles also have a lot of the enzyme lactate dehydrogenase, which removes the painful by-product of anaerobic respiration, lactic acid. Lactic acid causes muscle cramps. Species with the greatest diving ability have a higher proportion of lactate dehydrogenase in their muscles.

▼ **Emperor penguin**
Although the muscles in the wings are not large, those that work the wings as the bird "flies" through water are powerful.

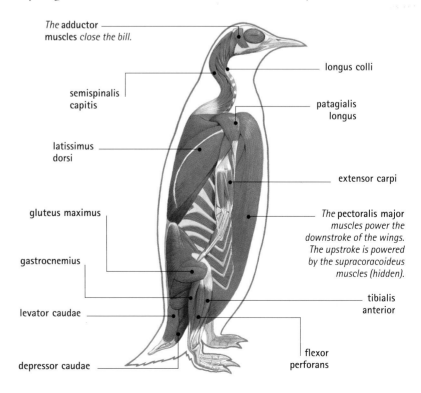

The adductor muscles *close the bill.*

semispinalis capitis

longus colli

patagialis longus

latissimus dorsi

extensor carpi

gluteus maximus

The pectoralis major *muscles power the downstroke of the wings. The upstroke is powered by the supracoracoideus muscles (hidden).*

gastrocnemius

levator caudae

tibialis anterior

flexor perforans

depressor caudae

Nervous system

Many birds, including penguins, are capable of learning new tasks and are relatively adaptable to changing circumstances. Penguins compare favorably with most mammals in overall learning ability, and trainers in zoos are able to teach the birds to perform simple tricks for a reward of fish.

Relative to their reptilian ancestors, birds have a large brain. Birds' brains are larger partly because birds are so reliant on eyesight—perhaps more so than any other animals apart from primates. Penguins use their eyesight to catch fast-moving prey and to communicate visually. Like other birds, a penguin's big brain, and particularly its large optic lobe, enables the penguin to process large amounts of visual information fast.

Ultra-sensitive eyes

When hunting, penguins must dive and catch their prey by sight. Penguin eyes are particularly sensitive to the dim blue-green light of the temperate and polar oceans. Only a little red light penetrates the plankton-rich water, so a penguin's visual pigments are biased in their color sensitivity toward the blue and violet end of the spectrum. Penguins can even see a little ultraviolet light but cannot see far into the ultraviolet spectrum, as can songbirds and parrots. Several penguin species, such as the king, emperor, and gentoo penguins, dive deeper than 150 feet (50 m) and hunt in virtual darkness. It is a mystery how they find their food at such depths, but it could be by seeing a biologically produced light, called bioluminescence, given off by the prey.

Although, like most other birds' eyes, penguins' eyes look sideways rather than forward, penguins see a 40° arc of space in front of them with both eyes—that is, they have a partly binocular visual field. Binocular vision is useful for judging distance and homing in on prey.

Penguins, like all birds, have similar ears to mammals, but instead of the three mammalian middle ear bones called ossicles, they have only one, the stapes. In mammals and birds,

IN FOCUS

Underwater vision

Penguins can see equally well in both air and water, owing to the structure of the lens and cornea in each eye. Land vertebrates rely on their curved cornea, as well as their lens, to focus light onto the retina. However, a cornea cannot focus in water because there is hardly any difference in refractive index (the degree to which light waves are "bent" as they pass from one medium to another) between water and the fluid inside the eye. Penguins therefore have flat corneas, which minimize differences between performance in air and water. Large differences between air and water remain, however, and penguins cope by radically changing the shape of their lens with muscles within the eye—from thin and flat in air to fat and round in water.

the ossicles' function is to amplify and transfer sound vibrations from the ear drum into the fluid of the middle ear.

Penguins rely on sound to recognize the voices of their mates and chicks in the large, dense breeding colonies typical of these birds. Species that build nests or burrows, such as African penguins and others in the *Pygoscelis* genus, recognize either the time structure (rhythm) or the harmonic structure (tone) of the calls. However, king and emperor penguins do not have nests or meeting sites, and their mechanism for distinguishing sounds is more sophisticated. They combine both rhythm and tone information to recognize the calls of mates and chicks.

Smell also seems to be important to penguins, since they have a large olfactory lobe, that part of the brain responsible for processing smell information. Most birds have small olfactory lobes, but many seabirds, such as albatrosses, shearwaters, and petrels, have a large olfactory lobe and a keen sense of smell.

Circulatory and respiratory systems

The respiratory and circulatory systems work together to exchange oxygen and carbon dioxide between a penguin and the air, and to pass these gases quickly around the bird's body. The circulatory system also transports other substances around the body and maintains the penguin's body temperature of around 100°F (37.8°C) in a huge range of air and water temperatures. Humboldt and Galápagos penguins that live near the equator experience air temperatures of more than 86°F (30°C); emperor penguins endure the winter exposed on the Antarctic ice cap in temperatures as low as −76°F (−60°C).

Miraculous net

Penguins are extremely well insulated from their environment by their feathers and by a layer of fat beneath their skin; this fat reaches a thickness of more than 1 inch (2.5 cm) in emperor penguins. Penguins have exposed areas, however, with little or no protective feathering: their feet, their flippers (especially the featherless undersides), and, in some species, most of the face. These areas are where the penguin's circulation helps conserve heat. The blood vessels that supply the bare areas are arranged in an ingenious way. The arteries

carrying warm blood from deep inside the penguin's body run very close to the veins carrying cooler blood inward from the feet and flippers, so the arterial blood is cooled on its way out to the extremities. The cool incoming blood is warmed in return, and the penguin conserves heat deep in its core, while allowing its feet and flippers to cool by up to

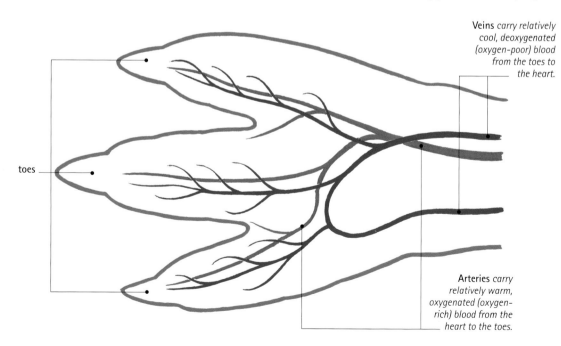

Veins *carry relatively cool, deoxygenated (oxygen-poor) blood from the toes to the heart.*

toes

Arteries *carry relatively warm, oxygenated (oxygen-rich) blood from the heart to the toes.*

◀ COUNTERCURRENT HEAT-EXCHANGE MECHANISM
Penguin's foot
Arteries carrying warm blood from the bird's core pass close to veins carrying cool blood from the extremities. Thus the venous blood is warmed.

▲ Emperor penguins swim in near-freezing ocean water. A feature of their circulatory system called the countercurrent heat exchange mechanism helps them retain heat in this harsh environment.

16°F (9°C) relative to the core. This heat-conserving arrangement is called a counter-current heat exchange mechanism.

In the head and face there is not enough room for countercurrent heat exchange to occur in the same way. Instead, the heat-exchanging veins and arteries divide into a fine mesh called a rete mirabile, or "miraculous net." The mesh has a large enough surface area to achieve the heat exchange to and from the naked face.

There is also a countercurrent heat exchanger in the respiratory system. Air from the lungs mixes with incoming air in a common chamber in the nasal passages before it is exhaled from the body. The nasal lining absorbs the excess heat and moisture. When the air from outside is at 41°F (5°C), the outgoing air can cool from 100°F (38°C) to 48°F (9°C), reclaiming 82 percent of the heat and water added to the air while it was inside the bird's respiratory system.

Holding their breath

The respiratory and circulatory systems of penguins allow the birds to be excellent divers. While diving, penguins cannot take in fresh oxygen, so they must function on stored oxygen, holding their breath. Penguins must also cope with other hazards of diving, such as the cold and the great pressure at depth, decompression when rising to the surface ("the bends"), and nitrogen narcosis (the toxic effect of dissolved nitrogen in the blood).

Penguins maximize their oxygen storage by several methods. The concentration of hemoglobin in penguin blood is high. Penguin hemoglobin binds with oxygen only loosely, so it readily releases oxygen into the tissues where it is needed. Penguins also have a huge oxygen store in the myoglobin of their muscles and in their lungs and air sacs. Penguins can reduce their need for oxygen during diving by selectively cooling their abdomen, which contains their digestive system; so penguins do not digest food when diving. When the abdominal organs are cold, they do not demand so much oxygen.

Acid buildup

Many experts estimate that, despite these adaptations, penguins run out of oxygen during dives that last more than a few minutes. To release energy, penguins would then have to resort to anaerobic respiration, which is much less efficient but does not need oxygen. The waste from anaerobic respiration is lactic acid, which becomes harmful when it builds up in the muscles. Penguin blood contains many ions that neutralize lactic acid, and the blood also contains the enzyme lactate dehydrogenase, which breaks down lactic acid. The penguin species that are best at diving have the most lactate dehydrogenase.

Oxygen must reach all the penguin's tissues, where it is used to release the chemical energy in the bird's food by the chemical process of cellular respiration. Likewise, carbon dioxide, the waste product of the penguin's respiration, must be expelled quickly from the body before it builds up.

Most birds keep their body temperature several degrees higher than that of mammals, at around 104°F (40°C), but penguins' body temperature is lower than that of most birds, at around 100°F (37.8°C), to keep it nearer the outside temperature. Penguins are so well insulated that even Antarctic penguins can maintain body temperature on land without generating extra heat with their body chemistry.

Digestive and excretory systems

A penguin grasps prey with its bill, a bony structure covered with horny keratin. (Keratin is the key structural protein in human hair and fingernails.) The keratin wears down and is replaced throughout the penguin's life. The slippery prey a penguin seizes does not often escape. The penguin grips its prey with its tongue, which is covered with backward-pointing spines called lingual papillae. Mammals have a soft palate at the top of the mouth, which helps them swallow, but penguins and other birds lack this feature; they must throw their head back to move food into the esophagus, or throat.

Since the esophagus is wide and muscular, it is able to force large, whole prey down to the stomach. Many birds have an enlargement of the esophagus called a crop, where they store food, often for their chicks. Penguins do not have a crop but instead regurgitate undigested food for their chicks directly from their stomach. Emperor penguins are exceptional, since the males produce a kind of "milk" from their esophagus lining (as do pigeons and doves), with which they feed their chicks.

Penguins, like other birds, have a two-part stomach. The first part, the proventriculum, has a glandular lining that produces mucus, acid, and enzymes that break down the proteins in food. King penguins halt this acidic digestion for up to three weeks when incubating chicks. In this way, they have undigested food in their stomach to feed their chicks if their mate does not return with fresh food.

The second part of the stomach, the gizzard, contains gastroliths (stones that the birds have swallowed) and is muscular with an abrasive, horny lining. The penguin's gizzard churns and breaks up its food. Some experts believe penguins use gastroliths for buoyancy, not digestion, because the fish, squid, and plankton food of penguins does not need much breaking down. Gastroliths are more common in seed-eating birds than in meat eaters.

COMPARE the stomach of a penguin with that of a **SEAL**, another animal that "flies" underwater. The stomach of both animals contains gastroliths, stones that probably help control the animal's buoyancy.

CONNECTIONS

IN FOCUS

Salt glands

Penguins drink and absorb seawater all the time. The water contains far more salt than is healthy for penguins, and so it must be removed from the body. As blood passes through a penguin's kidneys, they filter out as much salt as possible, but they cannot cope with the workload. Penguins also have salt glands, called supraorbital glands, which open into the nose. The glands concentrate salt from the penguin's body fluids and excrete it in a concentrated solution that dribbles down the bill and sprays when the penguin shakes its head.

▶ **Emperor penguin**
The mouth has few mucus or salivary glands. Food slips down easily and does not need lubrication. The esophagus is wide and muscular, and there is no crop. The small and large intestines of penguins are similar to those of mammals, but the penguin's end in a common opening to the outside called the cloaca. This opening is shared by the urinary and reproductive systems.

The cloaca is the common opening of the urinary, reproductive, and digestive systems. It receives uric acid from the kidneys, feces from the large intestine, and sperm or eggs from the gonads.

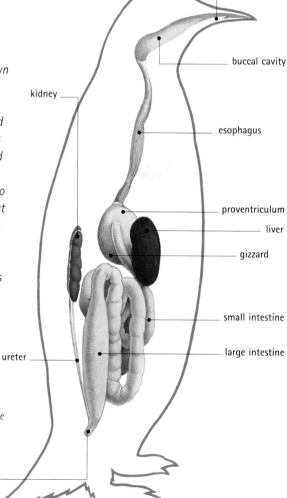

The tongue is covered with backward-pointing, spiny lingual papillae. There are few taste buds.

buccal cavity

esophagus

kidney

proventriculum

liver

gizzard

small intestine

large intestine

ureter

Reproductive system

The exit to a penguin's reproductive system, identical in both males and females, is a simple opening called the cloaca. Since the male has no penis (in common with most birds), he passes sperm to the female by a brief "cloacal kiss" lasting only a few seconds. To bring the male's cloaca into contact with the female's, the couple must perform an act something like balancing one football on top of another—it requires cooperation and synchronization. If sperm are successfully passed to the female, they may fertilize one of the female's eggs as it passes down the oviduct.

The oviduct is the tube connecting the ovary, where eggs are produced, with the cloaca and the outside world. Penguins, like most birds (but unlike mammals), have a single ovary and only one oviduct. In Fiordland crested and Adélie penguins, the yolk develops inside the ovary for 14 to 17 days; the egg is then released from the ovary and starts its journey down the oviduct. This process is called ovulation.

The oviduct of a penguin, like that of other birds, has five parts. The infundibulum is a funnel at the beginning of the oviduct. Glands in its epithelium, or lining, lay the first, dense layer of albumen (egg white) onto the egg. The magnum is the longest and most coiled section of the oviduct. It has a thicker wall and deeper glands, and it coats the egg with more albumin (egg-white protein). In the isthmus, the bird secretes the egg membranes, which

▼ **EGG BROODING**
King penguin
In the summer, female king penguins lay an egg, which is incubated for 51 to 57 days. This species of penguin does not build a nest. The male or female parent walks around with the egg resting on the top of the feet.

▼ **EGG POSITION**
King penguin
Emperor and king penguins have a naked patch of skin on their underside called a brood patch. During the incubation of an egg, this patch is swollen and filled with tiny blood vessels that can transfer heat to the egg.

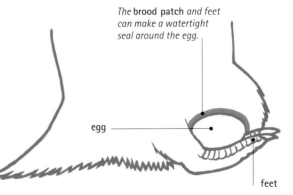

*The **brood patch** and feet can make a watertight seal around the egg.*

egg

feet

surround the egg. The next section is called the uterus, or shell gland uterus. There, water is added to the egg, and it doubles in weight. Then the shell gland secretes calcium, which coats the egg with its hard, mineralized shell. Finally, the muscular vagina pushes the egg out toward the cloaca. At the last moment, the vagina is pushed a little inside-out by the egg, and the egg clears the bird's body, at the same time pushing closed the exits to the reproductive and urinary systems, so the egg remains clean.

A basic nest

Few penguin species build elaborate nests, and most consist of little more than a few sticks and stones. Perhaps for that reason, penguins lay eggs with very hard shells. Those of the Magellanic penguin of South America are more than 50 percent thicker than shells of eggs of a similar mass laid by other bird species. Female Magellanic penguins actively select whole shellfish in their diet before egg laying. This way, they can store enough calcium in their body to lay down the thick eggshells.

Emperor and king penguins have no nest at all and support the egg on their feet. Like

▲ *An adult emperor penguin with its chick. The male parent incubates the newly laid egg while his partner goes to sea to feed. The female returns with food around the time the egg hatches. If the egg hatches before the female returns, the male is able to regurgitate a curdlike substance with which he feeds the chick.*

IN FOCUS

Feather phases

When penguin chicks are born they are altricial; that is, they are still at an early stage of development and are helpless. They have a downy covering called protoptile plumage. It is sparse and insufficient for insulation, but it allows the parent's body heat to transmit quickly through to the chick's body as the parent broods the chick. Toward the end of the brooding period, the chick molts to a thick, fluffy plumage called mesoptile plumage. This covering can insulate the chick while both parents leave together to forage. This is the peak foraging period, because the large chick has a huge appetite. Mesoptile plumage is not waterproof and is ineffective when wet, so chicks that have not molted to their waterproof adult plumage at the end of the breeding season are left behind as the colony enters the sea.

many other birds, on their underside they have a naked patch of skin called a brood patch. During the brooding period, the brood patch becomes swollen and filled with tiny blood vessels that can transfer body heat quickly to the egg or chick. The brood patch can be closed and watertight during swimming.

ROB HOUSTON

FURTHER READING AND RESEARCH
Lynch, Wayne. 1997. *Penguins of the World.* Firefly Books: Toronto.
Proctor, N. S., and Patrick J. Lynch. 1993. *Manual of Ornithology.* Yale University Press: New Haven, CT.
Schafer, Kevin. 2000. *Penguin Planet: Their World, Our World.* Northword Press: Chanhassen, MN.

Platypus

ORDER: Monotremata FAMILY: Ornithorhynchidae
GENUS: *Ornithorhynchus*

The platypus lives in and around the streams and lakes of eastern and southeastern Australia, where it feeds mainly on underwater plants and aquatic invertebrates, especially insect larvae, freshwater shrimp, and small mollusks. The unusual anatomy of the platypus incorporates characteristics of reptiles as well as mammals, together with highly specialized features associated with the platypus's unique method of detecting prey.

Anatomy and taxonomy

Scientists categorize all organisms into taxonomic groups based on anatomical, biochemical, and genetic similarities and differences. The single species of platypus and the four species of echidnas are members of the mammalian order Monotremata. These are the only mammals that lay eggs.

● **Animals** The platypus, like other animals, is multicellular and gets its food by consuming other organisms. Animals also differ from other multicellular life-forms in their ability to move from one place to another (in most cases, using muscles). They generally react rapidly to touch, light, and other stimuli.

● **Chordates** At some time in its life cycle, a chordate has a stiff, dorsal (back) supporting rod called the notochord that runs all or most of the length of the body.

● **Vertebrates** In vertebrates, the notochord develops into a backbone (spine or vertebral column) made of units called vertebrae. The muscular system that moves the head, trunk, and limbs of vertebrates consists primarily of muscles in a mirror-image arrangement on either side of the backbone (bilateral symmetry about the skeletal axis).

● **Mammals** Mammals are warm-blooded vertebrates that have hair made of keratin. Females have mammary glands that produce milk to feed their young. In mammals, each side of the lower jaw is a single bone, the dentary, hinged directly to the skull—a different arrangement from that found in other vertebrates. A mammal's middle ear contains three small bones (ear ossicles), two of which came from the jaw mechanism in mammalian ancestors. Mammalian red blood cells, when mature, lack a nucleus; all other vertebrates have red blood cells that contain a nucleus.

● **Nonplacental mammals** This group includes all monotremes and some marsupials. Nonplacental mammals do not nourish their unborn young through a placenta.

▼ *The monotremes are egg-laying mammals. They form one of the three major mammalian groups; the other two are the eutherians (placental mammals) and the marsupials (mammals whose young develop in a pouch). Today, only five species of monotremes exist, though several fossils of extinct monotremes have been discovered.*

Animals
KINGDOM Animalia

Chordates
PHYLUM Chordata

Vertebrates
SUBPHYLUM Vertebrata (or Craniata)

Mammals
CLASS Mammalia

Monotremes
ORDER Monotremata

Platypus
FAMILY Ornithorhynchidae

Echidnas
FAMILY Tachyglossidae

Platypus
GENUS AND SPECIES
Ornithorhynchus anatinus

Short-beaked echidna
GENUS AND SPECIES
Tachyglossus aculeatus

Long-beaked echidnas
GENUS
Zaglossus

Instead, nonplacental mammals either lay eggs (as in monotremes) or give birth to very undeveloped young that complete their development in their mother's pouch (as in marsupials).

● **Monotremes** The word *monotreme* means "single-holed" and refers to the cloaca, the common chamber into which the digestive, excretory, and reproductive systems empty. In other mammals, the digestive system empties to the outside separately from the urinary and reproductive systems. Male monotremes have spurs on their rear heels, although these are retracted and nonfunctional in echidnas. Monotremes are true mammals, with several characteristics unique to mammals. These features include a single bone called the dentary on either side of the lower jaw, three bones in the middle ear (the incus, stapes, and malleus), and milk-producing mammary glands. Unlike placental mammals and marsupials, however, monotremes' milk is secreted directly onto a patch of the body surface rather than through teats. Like other mammals, monotremes are endothermic: they regulate their body temperature by controlling internal heat production, as well as heat loss and gain across the body's surface.

● **Echidnas** The echidnas, sometimes called spiny anteaters, are the four members of the family Tachyglossidae. The name comes from *tachy,* meaning "fast," and *glossus,* meaning "tongue." It refers to the long, flexible tongue that flicks out at lightning speed to capture prey. Echidnas are covered with spines reminiscent of those of a hedgehog or porcupine. Echidnas have a long, narrow snout used for probing for food, and powerful claws for digging. The short-beaked echidna, widespread in Australia and Papua

▲ *Although the form of the platypus may seem unusual, its anatomical features are well suited for an aquatic lifestyle in which food is obtained by foraging on the beds of rivers and lakes.*

New Guinea, specializes in eating ants and termites. The three long-beaked species of the highlands of Papua New Guinea favor earthworms and larger insects.

● **Platypus** The platypus is the only living member of the family Ornithorhynchidae. The name *platypus* means "wide feet" and *ornithorhynchus* means "bird-snouted." The platypus is suited for life partly in water and partly on land, and its rubbery bill is a unique organ used for detecting small invertebrate prey that live in streams. The platypus is covered with a coat of soft, fine fur and larger guard hairs that together provide effective insulation in water. The feet are partially or completely webbed, and act as paddles in the water. The fur-covered tail is shaped like that of a beaver and is used for steering.

External anatomy

CONNECTIONS

COMPARE the limbs of a platypus with those of an *OTTER*. In otters, the hind limbs rather than the forelimbs are fully webbed. They work with the tail to propel the otter through the water when it is swimming at speed.

At first sight, a platypus looks rather as if it has been assembled from the parts of other animals—the bill of a duck, the body of an otter, and the tail of a beaver—but these similarities are superficial. The bill is similar in shape to that of a duck, but has a different structure and function. Like ducks, however, adult platypuses do not have teeth. The young have small teeth with stubby roots but these are replaced by hornlike plates on both jaws as the animal grows. The plates have sharp ridges at the front of the mouth for grasping prey but become flat toward the back for crushing food before it is swallowed. A pouch within each cheek is used to store food while the platypus is diving. Chewing and swallowing take place at the water's surface.

Insulating fur

The fur-covered body of the platypus is superficially similar to that of an otter, although the platypus differs from otters anatomically in many ways. Platypuses have a combination of features for life both in and out of water. The platypus's fur is dense and water-repellent, and the body shape is streamlined for swimming underwater. In common with the hair of otters, bladelike protective guard hairs overlie

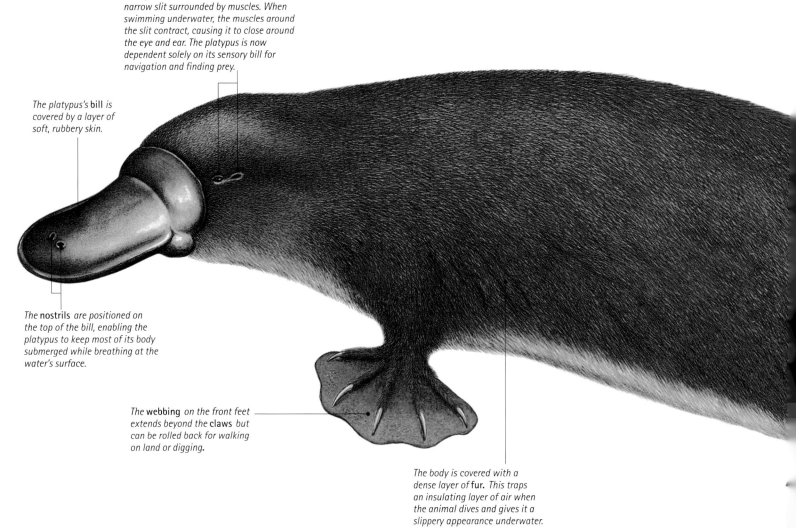

The **eye** and **ear** are contained within a narrow slit surrounded by muscles. When swimming underwater, the muscles around the slit contract, causing it to close around the eye and ear. The platypus is now dependent solely on its sensory bill for navigation and finding prey.

The platypus's **bill** is covered by a layer of soft, rubbery skin.

The **nostrils** are positioned on the top of the bill, enabling the platypus to keep most of its body submerged while breathing at the water's surface.

The **webbing** on the front feet extends beyond the **claws** but can be rolled back for walking on land or digging.

The body is covered with a dense layer of **fur**. This traps an insulating layer of air when the animal dives and gives it a slippery appearance underwater.

EVOLUTION

Monotremes and marsupials

The monotremes, bats, and marsupials, such as kangaroos, koalas, and wombats, are the only mammals with a history in Australia going back more than 15 million years. About that time a small group of rodents, the "old invaders," made it to the continent; but other types of land-living placental mammals (eutherians) became widespread in Australia only in the last few tens of thousands of years, beginning with the dingo. The dingo evolved from the domestic dog and was probably introduced by aboriginal people who arrived in Australia from Papua New Guinea within the last 65,000 years.

Because of the long coexistence of monotremes and marsupials, some anatomists have claimed that the two groups are closely related. However, the balance of evidence—from comparing anatomy, genetics, and biochemistry—suggests that marsupials are more closely related to eutherian (placental) mammals than they are to monotremes.

▼ *The platypus is instantly recognizable by its distinctive ducklike bill, furry body, and paddlelike tail.*

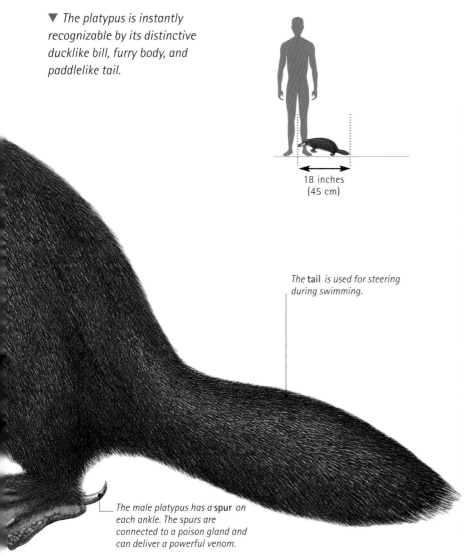

18 inches
(45 cm)

The tail is used for steering during swimming.

The male platypus has a **spur** on each ankle. The spurs are connected to a poison gland and can deliver a powerful venom.

fine underfur that traps air for insulation. The guard hairs are fairly stiff and spring upward to help lift the underfur. This enables the fur to dry rapidly and assists in the entry of air between the hairs for insulation. The density of underfur hair is about 450,000 hairs per square inch (70,000 hairs per cm²), which is close to that of some otters. Fur color in the platypus ranges from dark brown above to yellowish brown beneath and provides some camouflage for the animal at times when it is moving about in the open: at dusk and dawn and during the night. The tight-fitting burrows in which the platypus stays during the day and regularly retreats to at night, probably help squeeze water out of the fur, with the soil absorbing water like a sponge.

Limbs

Platypus feet are five-clawed. The forefeet are fully webbed with broad nails, but the hind feet are partially webbed and have sharp claws. The forelimbs are more important in swimming and digging. The hind limbs help the platypus steer

IN FOCUS

Strange but true

When the first platypus specimen (a dried skin) was brought back to England from Australia in the 1790s, it was thought to be a hoax. Scientists suspected that the bill of an unusual duck had been skillfully sewn onto the skin of an otter. Decades later, more complete specimens arrived and scientists began to accept the strange creature as real. But was it a mammal? Although the platypus had hair, its reproductive system was more similar to that of a reptile or a bird than a mammal, and it laid eggs. Other features were also reptilelike. The platypus's shoulder girdle was arranged like that of a therapsid, a type of small dinosaur; and platypus sperm were threadlike, like those of reptiles, rather than tadpolelike as in other mammals. Gradually, however, the case for the platypus being a mammal became overwhelming. The strongest evidence was that the platypus feeds its young with milk from mammary glands— something only mammals do.

▲ **Short-beaked echidna**
The short-beaked echidna's long spines provide it with some protection from predators. The long nose is used to probe into narrow spaces for ants and termites, which are drawn into the mouth by the long sticky tongue.

EVOLUTION

Platypus ancestors

The earliest platypuslike mammal, called *Steropodon*, dates back some 120 million years, to the Cretaceous period during the age of dinosaurs. Judging by its fossilized lower jaw, *Steropodon* was nearly twice as big as the modern platypus and had a bill armed with teeth. It was probably one of the largest mammals living at the time.

in water and provide firm anchors during digging. The platypus tail is broad and flat and acts as a rudder for steering underwater.

The pair of nostrils on top of the bill enable the platypus to breathe by just poking its bill slightly out of the water. The eyes and ears of the platypus are small and situated in a muscular groove on either side of the head. Although the senses of sight and hearing are useful when the animal is at the water surface or climbing on the bank of a stream, the sides of the muscular groove draw together when

the animal dives, closing both the eyes and the ears. Then, the animal's bill, which is soft and pliable and covered by dark skin, becomes essential for navigation.

Sensory bill

The platypus's bill is studded with pores that contain two types of sensory cell. Some of the cells are very sensitive to touch and to changes in water pressure and are able to detect the disturbances produced by moving prey. Other sensory cells in the bill detect the electrical fields produced by the muscular activity of prey. Using these senses, the platypus can find

◀ **BILL**
Platypus
The bill contains thousands of tiny sensory pits that detect water movement and electrical fields.

CLOSE-UP

The venomous spur

The venom gland of the male platypus grows larger in the breeding season and males become more aggressive at this time. The spurs are probably used as weapons to settle disputes between males. They fight over territory in which they can attract and keep a mate. Male echidnas have remnants of spurs, but they are hidden under the skin and do not release venom. This suggests that the spurs of echidnas might have functioned in their distant ancestors but have fallen into disuse.

Platypus venom contains more than 25 active ingredients. Some increase blood flow, causing swelling at the wound site; others lower the victim's blood pressure, causing shock. Various digestive-enzyme components dissolve tissues in the vicinity of the stab wound, causing the venom to spread inside the victim. One protein component specifically targets pain receptors, causing excruciating pain.

▶ **Internal view of spur**
The venom is squeezed out of the gland and into the reservoir by the surrounding muscle. It then passes through the spur and into the victim.

▼ **Exposed spur**
The male platypus uses his spur as a defense against attack and for competing with other males for reproductive rights.

venom gland

muscle

reservoir

spur

hind foot

▶ *The platypus's fur is dense and water-repellent, enabling it to retain body heat while swimming in water. Monotremes maintain a body temperature of 86 to 90°F (30–32°C), which is several degrees cooler than is typical of other mammals.*

its way around by detecting turbulence in the water and the electrical fields of animals in its surroundings.

Venomous spur

The male platypus and a few species of shrews are the only venomous mammals. The adult male platypus has a hollow venomous spur projecting from the back of the ankle of each hind foot. If challenged by another male or protecting itself against a predator, the male platypus can erect the spur and stab it into the attacker. Venom from a gland in the playtpus's thigh fills a reservoir inside the hollow spur. When the venom flows into a human victim, it causes agonizing pain. The venom is potent enough to kill a medium-sized dog.

Skeletal system

CONNECTIONS

COMPARE the positions of the upper limb bones in an platypus with those in an *OTTER*.

COMPARE the rib cage of a platypus with that of a reptile such as a *JACKSON'S CHAMELEON*.

▼ The platypus's skeleton has features that are common to mammals and reptiles. The jawbones and ear bones are typically mammalian, but some of the ribs connect to neck vertebrae, a feature more commonly found in reptiles.

As in other vertebrates, the platypus's skeleton shapes and supports the body; protects vital internal organs such as the brain, heart, and lungs; and serves as a point of attachment for skeletal muscles that move parts of the body. The skeleton of the platypus and the echidnas shows some typically reptilian features, particularly in the arrangement of some of the bones of the shoulder girdle and skull. However, the skeleton also has some classic mammalian features. One is that the lower jaw is made up of a single bone, the dentary, on each side, not several bones as in reptiles. Another mammalian feature is the three middle ear bones, whereas reptiles have just one.

Skull and jaw

In the three species of monotreme, the front of the skull forms the long snout, or rostrum, which is covered with soft, rubbery, sensitive skin with a rich blood supply. The platypus's rostrum is supported by widely spaced premaxillary and maxillary bones in the upper jaw and thin dentary bones in the lower. In the echidnas, these structures are much closer together, producing the echidna's characteristic narrow pencil-like snout. The monotreme cranium, or braincase, unlike that of most other mammals, is smoothly rounded. The sutures—the places where the constituent bones have fused—are smoothed over, obscuring the joins. Scientists, therefore, have difficulty in correctly interpreting the skull bones of monotremes in relation to other mammals. The sides of the monotreme braincase, for example, are made up of petrosal bones; in all other mammals, alisphenoid bones perform this role.

Echidnas do not have teeth. Young platypuses have some small nonfunctional teeth. By the time young platypuses are weaned at four to six months, the teeth have been replaced by horny plates. The platypus's bill is larger on top than on the bottom and has a frontal shield that extends slightly up and over the forehead.

Limbs and their supports

The limbs of monotremes are connected to the spine through two limb girdles that contain a blend of reptilian and mammalian features. The pectoral (shoulder) girdle contains coracoid bones and a **T**-shape interclavicle that are typically reptilian. These bones help brace and support the legs, which are splayed out from

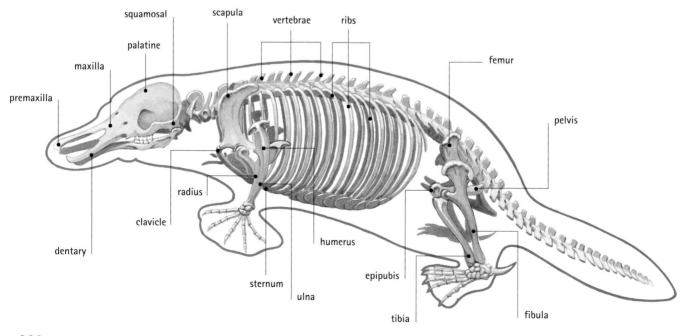

the main axis of the body. In monotremes, the upper leg bones (the humerus of the forelimbs and the femur of the hind limbs) are held roughly parallel to the ground, as in the reptilian ancestors of mammals, the therapsids, and in most modern reptiles. Placental mammals, in contrast, lack coracoids and interclavicles, and the legs extend beneath the body like supporting columns.

Monotremes have the scapula (shoulder blade) and clavicle (collarbone) of mammals, but these elements are much more strongly attached to the vertebral column than is typical of other mammals. In monotremes, ribs are attached to neck (cervical) and chest (thoracic) vertebrae, as in some reptiles. In all other modern mammals, ribs are restricted to the chest region.

The pelvic girdle, which connects the hind limbs to the vertebral column, is also unusual in monotremes. Marsupials and monotremes have epipubic bones that curve forward from the pelvis. Scientists once thought these supported the marsupials' and echidnas' pouch, which protected the growing infant. However, males do not have pouches; neither do female platypuses or some marsupials. It is more likely that the epipubic bones were functional in monotremes' ancestors. The epipubic bones probably anchored the large abdominal muscles in therapsid reptiles from which both monotremes and marsupials evolved.

Leg bones

The legs of platypuses and both species of echidnas are short; none of the monotremes is a fast runner. Platypuses spend most of their time in water or in burrows. When they walk, they do so on their knuckles and tuck the webbing out of the way to expose their claws. Echidnas, by contrast, walk on their palms with the body much higher off the ground. On soft ground, a frightened echidna can burrow straight down into the soil within a matter of seconds. On hard ground, where burrowing is not possible, a threatened echidna rolls into a ball with its spines projecting to the outside for protection.

Unlike the bones of the upper limbs, which are oriented horizontally, the lower-limb elements—the radius and ulna in the forelimb

COMPARATIVE ANATOMY

Bills and beaks

The platypus bill and echidna beak, when compared with those of certain birds, provide good examples of convergent evolution, where unrelated species evolve similar-looking structures in response to comparable lifestyles. In the platypus and in ducks, the bill serves as a tool for poking about at the bottom of streams and ponds and turning over stones to reveal prey. In echidnas and birds like the New Zealand kiwi, the long beak serves for probing into deep crevices—such as those in termite mounds—to extract insects.

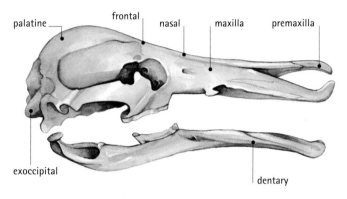

▲ SKULL (SIDE VIEW)
Platypus

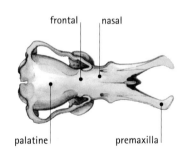

▲ SKULL (DORSAL VIEW)
Platypus
The premaxillae are widely separated at the front of the skull.

▼ SKULL (VENTRAL VIEW)
Platypus
The platypus's lower jaw is made of two dentary bones, a feature typical of mammals.

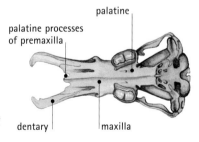

and the tibia and fibula in the hind limbs—are more or less vertical. These arrangements, together with the ways these bones are connected to the girdles through joints and ligaments, make the limbs rotate rather than simply move back and forth as they do in marsupials and placental mammals. Echidnas move on land with a curious rolling gait.

Muscular system

It contracts slowly to alter the volume of the space, or lumen, running through the tube. In the case of the gut, contracting smooth muscle squeezes food along the lumen.

Striated or striped muscle, so called because of its appearance under a microscope after staining, is the type of muscle attached to bone. By contracting, this type of muscle moves bones—hence its other name, skeletal muscle.

The largest skeletal muscles in the platypus are those in the upper part of the forelimbs. They power both swimming and digging movements. When swimming or digging, a platypus extends a forelimb on one side and then pulls it back, rotating it and swinging it in toward the body. The return stroke is made with the forelimb folded against the body until it reaches the start point once again. The muscles that produce the powerful pull stroke are retractors that pull the lower forelimbs backward, adductors that pull the forelimbs toward the body, and rotators, that rotate the limbs around the "elbow" joint.

When an adult male platypus is threatened, he may contract extensor (limb-straightening) muscles in the upper and lower hind limbs to kick backward and stab the attacker with the poison spurs. At the same time a mass of muscle around the base of the poison glands squeezes venom into the spurs from where it is delivered.

▲ *Powerful forearm muscles provide the thrust necessary for the platypus to swim underwater.*

Muscles make up more than one-third of the body weight of a platypus. They work by contracting to move parts of the body. Two kinds of muscle—smooth and striated—are widely distributed in the body, with a third type, called cardiac muscle, found in the heart.

Smooth muscle lies in the walls of tubular structures, such as the gut and major arteries.

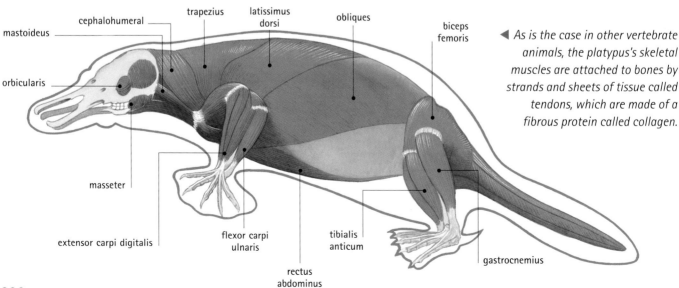

mastoideus
cephalohumeral
trapezius
latissimus dorsi
obliques
biceps femoris
orbicularis
masseter
extensor carpi digitalis
flexor carpi ulnaris
tibialis anticum
gastrocnemius
rectus abdominus

◀ *As is the case in other vertebrate animals, the platypus's skeletal muscles are attached to bones by strands and sheets of tissue called tendons, which are made of a fibrous protein called collagen.*

Nervous system

The brain of the platypus accounts for about 0.5 percent of its body weight, which is about the same proportion as in a rat, but considerably less than the proportion in a human (2 percent of body weight). The surface of the cerebral hemispheres in the platypus is smooth, with very little folding. This is normally an indication of a "primitive" brain with limited processing power. An echidna's brain, on the other hand, shows considerable cerebral folding. Such folding normally indicates greater processing power. This difference between the brains of the platypus and echidnas has yet to be satisfactorily explained.

The brains of monotremes exhibit a combination of reptilian and mammalian features. As in other mammals, the pallial, or upper, region is the most strongly developed part of the cerebral hemispheres. In reptiles and birds, the striatal, or lower, part is more pronounced. However, along with reptiles and marsupials, monotremes lack the corpus callosum, a thick band of nervous tissue that connects the right and left hemispheres of the brain in other mammals and integrates activity between the two sides of the brain.

Electroreceptors

The most remarkable feature of the platypus's nervous system is its ability to detect and home in on the electrical fields of prey. Within the

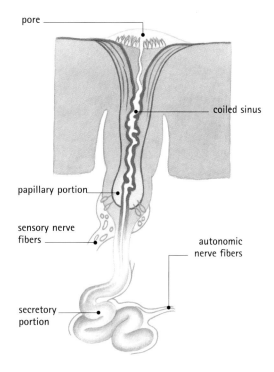

▲ SENSORY MUCUS RECEPTOR
The platypus detects electromagnetic fields using sensory mucus receptors on its bill.

▼ BRAINS
In many respects, the platypus's nervous system is similar to that of other mammals. A large proportion of the brain is devoted to processing sensory information from the bill.

COMPARE the electroreception system of the platypus with that found in the **HAMMERHEAD SHARK**.

COMPARE the platypus's system of "push rods," which sense touch or pressure, with the lateral line system of a bony fish such as a **TROUT**.

CONNECTIONS

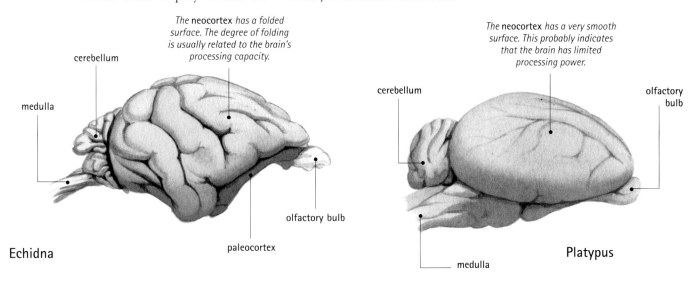

*The **neocortex** has a folded surface. The degree of folding is usually related to the brain's processing capacity.*

*The **neocortex** has a very smooth surface. This probably indicates that the brain has limited processing power.*

Echidna

Platypus

891

▶ *The platypus's bill is covered with thousands of sensory cells that detect changes in water pressure and the presence of electrical fields. The platypus uses these cells to navigate under water and to locate prey.*

EVOLUTION

The platypus's sixth sense

The platypus and echidnas are the only mammals with an obvious electrosensory system. This suggests that the monotremes branched off the main stem of mammalian evolution at an early stage and developed their unique sensory system independently of all other mammals. It probably took many millions of years to evolve the new receptors, nerve pathways, and brain processing machinery that produced the platypus's amazing "sixth sense." Other than monotremes, the only animals that have evolved a sophisticated electrosensory system are certain types of fish.

platypus bill are about 60,000 electrical receptors and about 40,000 touch or pressure receptors, sometimes called "push rods" because of their shape and the way they work.

The two types of receptors sit inside mucus-filled pits that open in water and close in air. These two types of receptor work together to pinpoint the precise location of underwater prey at ranges of a few inches. As a shrimp swims to escape, the muscles that move its tail produce a pulse of electricity that the platypus's electric receptors detect. A fraction of a second later, touch receptors in the platypus's bill sense the "wash" or pressure waves from the disturbance created by the moving shrimp. The time delay between the arrival of the electrical pulse and the physical disturbance is probably used by the platypus to give an indication of the prey's distance. This is similar to the way in which people can tell how far away a thunderstorm is by the difference in time between the flash of light and the sound of the thunderclap. The platypus can get an accurate fix on its prey, even though it cannot see, hear, or smell underwater.

Both touch and electrical information travel from the sensory receptors in the bill and connect to similar parts of the platypus's brain. In fact, some brain cells receive sensory information from both electrical and touch receptors. The platypus's electrical sense probably evolved from touch receptors that have now taken on a different role. Amazingly, about two-thirds of the sensory part of the platypus brain is connected to the bill.

Circulatory and respiratory systems

The platypus, like other diving mammals, does not breathe underwater. In order to supply tissues with oxygen during the dive, the mammal must take plenty of air down with it, or use other strategies to compensate for the lack of oxygen when diving. In practice, most diving mammals do not take large amounts of air down with them but solve the problem of lack of oxygen in various other ways. Expert divers such as seals, whales, and manatees reduce the rate at which the heart pumps blood, thereby reducing energy consumption and the need for oxygen. At the same time, some of the arteries constrict, blocking the flow of blood to less vital tissues and organs, while key organs such as the brain and heart continue to receive their supply of oxygen-carrying blood. A third method is the use of high levels of myoglobin (an oxygen-carrying pigment that is similar to hemoglobin) that occurs in muscle. The myoglobin traps oxygen that can be used during the dive. Last, most diving mammals have muscles that can respire (release energy from food substances) anaerobically—that is, without the need for oxygen.

Obtaining oxygen

Of these four strategies, platypuses use the first two but not the last two. At the surface, a

IN FOCUS

Diving

As well as short feeding dives that last 2 minutes at most, platypuses also make longer dives that last up to 10 minutes or more. In longer dives, the animal wedges itself under an object on the river or lake bottom and stays almost immobile for minutes at a time. The purpose of this behavior is not known.

platypus's heart beats between 140 and 230 times a minute. During a dive, it beats between 10 and 120 times a minute, and blood flow is restricted to only the most vital organs.

Unlike most diving mammals, platypuses take a good lungful of air with them when they dive. Like other diving mammals, they have blood that is rich in hemoglobin (the oxygen-carrying pigment) which allows their blood to carry large amounts of oxygen for use in diving. The hemoglobin is contained in red blood cells, which make up nearly half of the blood volume. Hemoglobin-rich blood also allows the platypus when it is in its burrow to extract plenty of oxygen from the air it inhales, even when the burrow air is fairly stale.

▼ *Like other mammals, the platypus has a four-chamber heart that provides the pumping power to force blood through the blood vessels. Respiratory gases, such as oxygen and carbon dioxide, enter and exit the blood across the surface of the lungs.*

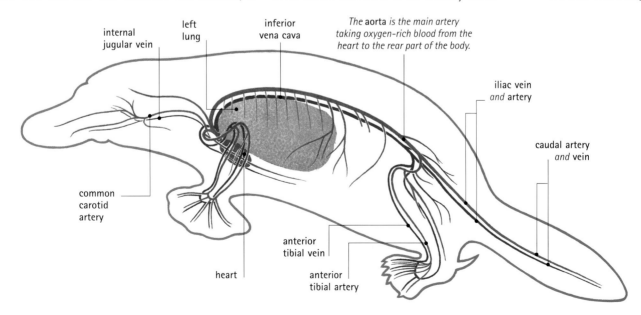

internal jugular vein

left lung

inferior vena cava

The aorta *is the main artery taking oxygen-rich blood from the heart to the rear part of the body.*

iliac vein *and* artery

caudal artery *and* vein

common carotid artery

anterior tibial vein

heart

anterior tibial artery

Digestive and excretory systems

Swimming above the riverbed, scanning for electrical fields with its bill moving from side to side, the platypus can find and consume one-third of its body weight in small invertebrates in a single night. It can gather several items of prey in a single dive, storing them temporarily in either cheek pouch, until it surfaces after a minute or two. On the surface, or in its burrow, it grinds the food to a pulp between the plates on its upper and lower jaws, while manipulating the food between its tongue and the roof of the mouth (palate) before swallowing it.

Platyfood

Platypuses are opportunistic feeders, taking a wide variety of prey animals depending upon what crosses their path. They consume crustaceans such as freshwater shrimp, mollusks such as mussels and freshwater snails, and a wide variety of aquatic insects both as larvae and as adults. The aquatic insects include caddis flies, dragonflies, mayfly larvae, swimming beetles, and water bugs. Platypuses will also take frogs, tadpoles, and small fish. They rarely, if ever, hunt on land but will take land animals that fall into the water. Platypuses require a large supply of food to maintain their active lifestyle and moderately high body temperature in a watery environment where they may lose a great deal of heat to their surroundings.

▲ *The platypus forages on the bed of lakes and rivers for invertebrate prey. Less often, small fish, frogs, and tadpoles are eaten.*

▼ *The platypus's intestine exits through the cloaca, which is the same opening that is used for excreting urine from the bladder and for reproduction.*

Although platypuses will eat water plants, these are too low in calories and difficult to digest to form a major part of the diet.

The intestines of a platypus are short relative to those of many other mammals, especially mammals that eat mostly plant material. When straightened, the intestines of a platypus are only about about three times its body length; this proportion is more typical of a flesh-eating mammal. The platypus has a relatively small stomach, which is typical for a meat eater that eats fairly continually rather than consuming large amounts at once.

▶ BILL CROSS SECTION
Cheek pouches provide temporary stores for food.

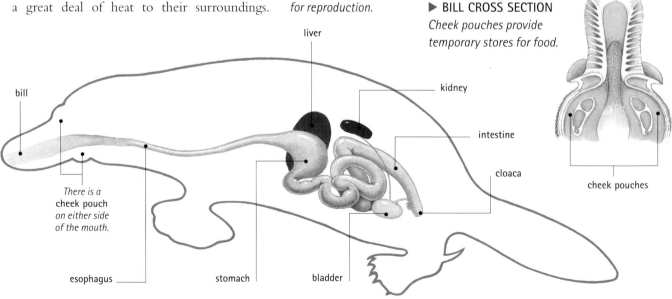

liver

kidney

intestine

cloaca

cheek pouches

bill

There is a cheek pouch on either side of the mouth.

esophagus

stomach

bladder

Reproductive system

In the breeding season in early spring, the male platypus courts the female by chasing her and grabbing her tail. Only the most dominant males that can defend a territory are successful breeders. The female responds to the male's behavior by grabbing his tail. The two spend much of their time swimming around, following each other in slow circles. Typically, after several days, they mate in the water. When mating, the male usually lies on top of the female and steers his erect penis (which is 2 to 3 inches, or 5 to 7.5 cm, long) into her cloaca where he releases sperm.

During the courtship period and in the few days that follow, the female excavates a burrow system larger than the one she or her partner usually occupies. The burrow may run for 10 feet (3 m) or more, with several chambers extending from it. The female lines one or more chambers with wet grass and other leaves, ferrying bundles of them tucked beneath her folded tail. The leaves help keep the eggs moist and prevent them from drying out during incubation.

Cloaca

The arrangement of the reproductive system in monotremes is different from that of all other mammals. In male and female platypuses and echidnas, the exiting duct from the reproductive system empties into a cloaca—a chamber into which both digestive and urinary tracts also empty. In all other mammals, urinary and reproductive tracts empty into a region separate from the digestive tract. Male monotremes produce sperm in two testes that

COMPARE the reproductive system of a female platypus with that of a female marsupial, such as a KANGAROO. Platypuses lay eggs but kangaroos give birth to live young. Both, however, have mammary glands with which they feed their young.

CONNECTIONS

Each glans terminates in a rosettelike structure, here shown unextended.

When the penis is erect, the rosettelike tips are thought to protrude like a flower.

The glans is divided into two asymmetrical parts, with the left side larger than the right.

The penis is covered with backward-facing spines made of a protein called keratin.

◄ PENIS
Platypus
The platypus penis is highly unusual. The tip of the penis (glans) divides asymmetrically into two parts, and each tip terminates in a flowerlike rosette. Sperm exit the penis through the urethra, which is composed of branching tubes. Each branch exits through the tip of one of the "petals."

▶ PENIS
Echidna
The glans of the echidna penis is also divided, but into four parts forming an overall symmetrical arrangement. In contrast with the platypus's penis, the echidna penis is smooth.

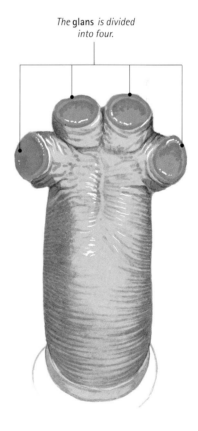

The glans is divided into four.

are retained within the abdomen in a manner similar to reptiles. Sperm leave each testis at the climax of mating, pass along the vas deferens to the urethra, and are squirted out through the unusual two-headed penis that emerges from the cloaca.

Monotreme females produce eggs in ovaries, which open separately into the cloaca through oviducts. As in birds, the platypus female has two ovaries, but the right one does not develop fully and does not produce eggs. In echidnas, both ovaries are fully functional.

In the platypus, fertilization occurs inside the functioning oviduct, or fallopian tube. The one or more fertilized eggs become covered with albumen (egg white) as they pass down the oviduct and through the left uterus (womb). Finally, a leathery shell is added. Monotremes incubate their eggs and hatch them outside their body, as birds do.

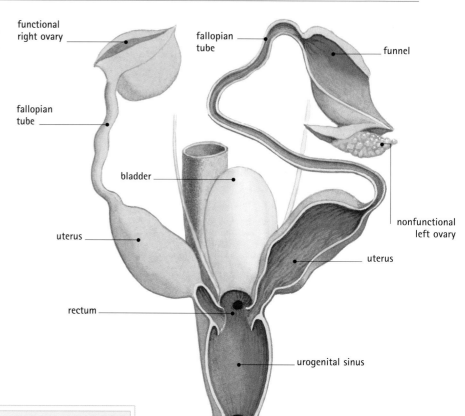

▲ FEMALE UROGENITAL SYSTEM, VENTRAL VIEW
Only the right ovary of the female platypus is functional. The cloaca provides a common exit chamber for urine, feces, and eggs.

The platypus's eggs are more like reptiles' eggs than birds' eggs, being sticky, leathery, and nearly spherical. The eggs hatch after 8 to 11 days. The young platypuses cut their way out of their shells using an egg tooth like that found in some reptiles and birds. The newborns are only about two-thirds of an inch (1.5 cm) long and are blind and naked.

Mammary glands

The mother curls around the young to keep them warm. She feeds them on milk that oozes from large mammary glands on her underside. She does not have teats, and the young simply lap up the milk from two soaked patches of fur midway along her belly.

IN FOCUS

Monotreme mammary glands

The mammary glands of monotremes differ from those of all other mammals. The glands, which are like modified sweat glands, are not collected together to form recognizable breasts, nor do they have teats. Instead, the milk is released directly from mammary gland ducts onto the surface of the skin. Young monotremes have a snout shaped to press against the surface of the skin and slurp the milk.

The echidna's pouch

Echidna females carry their single egg around with them, attached to the fur on their belly. After about 10 days, the young echidna cuts its way through the leathery shell and clambers several inches forward to a patch of hairs soaked in milk. At any time, the baby can withdraw into an incomplete and backward-facing pouch that lies just in front of the cloaca. By 10 days after hatching, the youngster is becoming spiny and uncomfortable to carry and the mother transfers it to a burrow. She returns regularly to feed her youngster milk. The young echidna is able to fend for itself after 8 to 9 months.

▼ *After laying a single egg, the female echidna carries it around on her abdomen. During the breeding season, she grows a rearward-opening pouch into which the hatched offspring can retreat.*

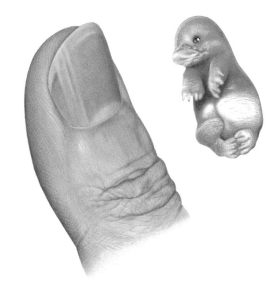

▲ *When an echidna hatches from its egg, it is tiny, measuring only about 0.5 inch (1.27 cm) and weighing 0.014 ounce (0.4 g).*

When the young press into the milk patches, this pressure triggers the release of a chemical called oxytocin, which encourages the mammary gland cells to contract, so squeezing out more milk onto the skin. Platypuses' milk is rich in fat, sugar, and protein, which together provide the young with much of the energy and raw materials they need to grow rapidly. The milk is also rich in iron. This mineral is needed for the young to manufacture hemoglobin so they can take up oxygen efficiently from the slightly stale air in the burrow.

Unlike female echidnas, the platypus mother does not have a pouch and does not carry her young with her when she leaves the burrow. She goes outside only occasionally to wash and wet her fur and defecate. When she comes out, she plugs the entrance to the burrow with loose soil to prevent easy access for predators.

The young emerge from the burrow, fully furred, after about four months, during which time they have grown rapidly to reach about 13 inches (33 cm) long. They quickly learn to swim and soon begin catching their own food, but probably do not become fully weaned until several weeks later. The young stay initially within the mother's home range, but move farther afield by the time they reach maturity at about two years old. In the wild, platypuses can live to about 12 years old, but most platypuses that reach adulthood live for 5 to 8 years, with females breeding once every two years.

TREVOR DAY

FURTHER READING AND RESEARCH

Grant, T. R. 1995. *The Platypus: A Unique Mammal* (2nd ed.) University of New South Wales Press: Sydney.

Rismiller, P. 1999. *The Echidna: Australia's Enigma.* Hugh Lauter Levin Associates: Westport, CT.

Porcupine

ORDER: Rodentia SUBORDER: Caviomorpha
FAMILY: Erethizontidae GENUS: *Erethizon*

The North American porcupine is one of 23 species of porcupines belonging to two families: the New World porcupines (Erethizontidae) and the Old World porcupines (Hystricidae). It is the only New World porcupine to occur in northern North America, where it is found throughout the forested regions of Canada and northern parts of the United States. The other New World species are restricted to tropical Central and South America. The Old World porcupines live in Africa, South Asia, and southern Europe.

Anatomy and taxonomy

All animals are classified in groups based mainly on shared anatomical features. These features usually indicate that the members of a group have the same ancestry, so the classification shows how the organisms are related to one another and to extinct fossil forms. In the case of porcupines, both families show the characteristics of cavylike rodents, but the spines, or quills, of New World and Old World species are different and probably evolved independently.

▼ *The porcupines belong to two families in the suborder Hystricognathi, the cavylike rodents. There are four genera of New World porcupines (*Coendou, Sphiggurus, Erethizon, *and* Echinoprocta*) and four genera of Old World porcupines (*Atherurus, Hystrix, Thecurus, *and* Trichys*).*

● **Animals** All true animals are multicellular organisms with well-developed powers of muscular movement and the ability to respond rapidly to stimuli. An animal obtains nutrients by eating other organisms and digesting their tissues. Its body uses these simpler molecules to provide energy or to build tissues.

● **Chordates** A chordate has a strong, flexible rod called a notochord running along its back. This supports its body and allows its muscles to work more effectively. Most chordates retain the notochord throughout life, but some simple groups, such as sea squirts, lose it as they mature.

● **Vertebrates** The notochord of a vertebrate forms the basis of a flexible backbone made up of units called vertebrae. The vertebrae and other skeletal bones provide anchorage for muscles that are mirrored on the left and right of the body (bilateral symmetry). A vertebrate also has a brain enclosed within a cranium, or skull, and the group is sometimes called the Craniata.

● **Mammals** Mammals are warm-blooded vertebrates that feed their young on milk produced by the females. Typical mammals have a covering of fur or hair, which is unique to the group. A mammal's lower jaw is hinged directly to its skull, unlike the jaws of all other vertebrates, and a mammal's red blood cells do not have a nucleus.

Animals
KINGDOM Animalia

Chordates
PHYLUM Chordata

Vertebrates
SUBPHYLUM Vertebrata

Mammals
CLASS Mammalia

Rodents
ORDER Rodentia

Cavylike rodents
SUBORDER Hystricognathi

Mouselike rodents and squirrel-like rodents
SUBORDER Sciurognathi

New World porcupines
FAMILY Erethizontidae

Cavies
FAMILY Caviidae

Capybara
FAMILY Hydrochaeridae

Chinchillas
FAMILY Chinchillidae

Old World porcupines
FAMILY Hystricidae

- **Placental mammals** Unlike marsupials and monotremes, placental mammals nourish their unborn young during pregnancy. Nutrients from the mother's blood are delivered through an umbilical cord and placenta, attached to the wall of the uterus.

- **Rodents** More than 42 percent of all mammal species are rodents. They are mostly small animals that feed mainly on tough plant material and are equipped with four large front (incisor) teeth that are used for gnawing. These teeth grow continuously to compensate for wear and are self-sharpening. There are two groups of rodents: cavylike rodents and squirrel-like and mouselike rodents.

- **Cavylike rodents** Named for the well-known cavies, or guinea pigs, cavylike rodents are typically plump with a relatively large head, and they include the largest of all rodents, the capybara. They share a distinctive pattern of jaw musculature, with the deep masseter muscle providing most of the gnawing power. They also bear relatively few young, which are well developed at birth.

- **New World porcupines** The New World porcupines are mainly arboreal (tree-living) and have strong claws and gripping feet for climbing; some species have a muscular, prehensile tail for grasping branches. Their quills grow singly from the skin, interspersed with long guard hairs. There are four genera of New World porcupines with 11 surviving species, including the North American porcupine.

▲ *The prehensile-tailed porcupines live in the rain forests of South America. They spend most of their time in the trees but occasionally come down to forage on the forest floor.*

- **Old World porcupines** The Old World porcupines live mainly on the ground, and although some feed in trees they are not especially suited to climbing. Their quills grow in clusters of four to six, rather than singly. Some species are able to rattle bunches of specialized quills to produce a warning rustle. Old World porcupines are usually grouped in four genera with 11 species, including the very spiny African porcupine, which also lives in southern Europe.

EXTERNAL ANATOMY Porcupines are spiny rodents with a variety of spine patterns and body forms, depending mainly on whether they live on the ground or in trees. *See pages 900–903.*

SKELETAL SYSTEM Porcupines have a short-legged skeleton and have teeth highly adapted for gnawing and chewing tough vegetation. *See pages 904–905.*

MUSCULAR SYSTEM Porcupines are adapted for a slow-moving lifestyle. Some tree-climbing species are equipped with a strong prehensile (grasping) tail for gripping branches. *See pages 906–907.*

NERVOUS SYSTEM Porcupines have a good memory and acute senses of smell, hearing, and touch. *See pages 908–909.*

CIRCULATORY AND RESPIRATORY SYSTEMS Oxygen passes through the lungs to the blood, which the heart pumps around the body. *See pages 910–911.*

DIGESTIVE AND EXCRETORY SYSTEMS The porcupine's long digestive tract contains cellulose-splitting bacteria for digesting fibrous plant material. *See pages 912–913.*

REPRODUCTIVE SYSTEM Porcupines produce one or two young at a time. Newborn porcupines are fully furred, with soft spines. *See pages 914–915.*

FEATURED SYSTEMS

External anatomy

CONNECTIONS

COMPARE the spiny quills of a porcupine with the tough shell of a *TORTOISE*. The tortoise's shell forms a defensive armor, but a porcupine's spines act as defensive weapons.

Porcupines are unusual-looking animals. The crested porcupines of Africa and South Asia are particularly striking, with their flamboyant crests of bristly hair and extremely long, sharp black-and-white spines. Some of the South American porcupines are equally remarkable, resembling animated brushes with beady black eyes and a squashed button nose. Of the 22 surviving species of porcupines, the least remarkable-looking is the long-tailed porcupine of Borneo, Sumatra, and the Malay peninsula. It resembles a bristly rat, with a brush of short spines on the tip of its scaly tail. The North American porcupine is a stout, somewhat scruffy-looking animal, with a tousled coat of long stiff hairs interspersed with short spines. This porcupine is a careful but accomplished climber.

Spines and rattles

The spines, or quills, are modified hairs. They are the defining feature of a porcupine, but in some species they are not immediately obvious. Those of the North American porcupine, for example, cover the back and tail but are usually disguised by longer, stiff guard hairs. The spines become apparent only when the animal bristles up its hair and spines in alarm. Otherwise, it looks more like a bristly, tree-climbing beaver. By contrast, the two species of brush-tailed porcupines, which live in Africa and Asia, are almost entirely spiny, although the spines on the underside of the body are softer and more hairlike than those elsewhere on the body.

The longest spines belong to crested species, such as the African porcupine. At up to 20 inches (50 cm) long, the spines provide a formidable

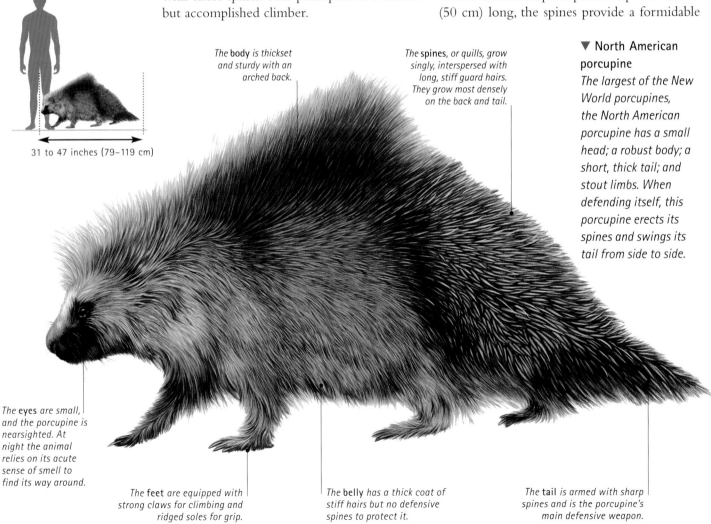

31 to 47 inches (79–119 cm)

*The **body** is thickset and sturdy with an arched back.*

*The **spines**, or quills, grow singly, interspersed with long, stiff guard hairs. They grow most densely on the back and tail.*

▼ **North American porcupine**
The largest of the New World porcupines, the North American porcupine has a small head; a robust body; a short, thick tail; and stout limbs. When defending itself, this porcupine erects its spines and swings its tail from side to side.

*The **eyes** are small, and the porcupine is nearsighted. At night the animal relies on its acute sense of smell to find its way around.*

*The **feet** are equipped with strong claws for climbing and ridged soles for grip.*

*The **belly** has a thick coat of stiff hairs but no defensive spines to protect it.*

*The **tail** is armed with sharp spines and is the porcupine's main defensive weapon.*

CLOSE-UP

Deadly spines

The 30,000 spines, or quills, of a North American porcupine are concentrated on its back and tail. If the porcupine feels threatened, it bristles up its spines and backs toward the assailant, shaking its tail with the aim of jabbing one or more of the extremely sharp spines into its enemy's skin. Each spine pulls out easily from the porcupine's skin, and the tip is covered with thousands of tiny backward-pointing barbs. If a spine penetrates an enemy, it lodges in the skin like a thorn. The regular tensing and relaxing of the victim's skin pulls on the barbs and draws the spine farther into the animal's flesh. It may be drawn right into the body and pierce a vital organ or cause a festering and possibly fatal wound. More usually, it is gradually destroyed by the animal's natural defenses, but by then it has done its job.

▼ SPINE
North American porcupine
Each spine ends in a sharp point and is covered with backward-pointing barbs. If a spine penetrates an assailant, the barbs keep it from slipping out, and in time it becomes deeply embedded.

barbs

EVOLUTION

Separate development

Old World and New World porcupines are all cavylike rodents, but they are not closely related. It is likely that over millions of years the Old World porcupines evolved their cavylike characteristics independently of the New World porcupines, through becoming adapted to cope with the same problems. The phenomenon of independently evolving similar features is called convergent evolution. The spines of the two groups of porcupines probably evolved independently because although they have the same function, their structures are different. The spines of New World porcupines grow singly, whereas those of Old World species grow in clusters, indicating that they have different origins.

spines with open-ended tubes at their tips. When the animals shake their tail, the hollow spines knock together to produce a hissing rattle, which porcupines use to threaten enemies in much the same way as a rattlesnake. The North American porcupine also shakes its tail in threat, but it has no rattle quills.

Crested heavyweights
The five species of crested porcupines of Africa and southern Asia are ground-dwelling animals, as are most of the Old World porcupines. Crested porcupines have short legs and a short tail and are plantigrade (they walk on the soles of their feet) with a heavy gait. They are the giants of the porcupine world, weighing up to 59 pounds (27 kg) when fully grown. The three

▼ *Old World porcupines such as this family of three crested porcupines have much longer spines than New World porcupines.*

defense when erected into a barbed fan—enough to deter powerful predators such as lions and leopards. At other times, the long spines lie flat over the animal's back, extending backward like a prickly mantle.

In addition to spines, crested and Indonesian porcupines are also equipped with specialized "rattle quills" on the tail. These are expanded

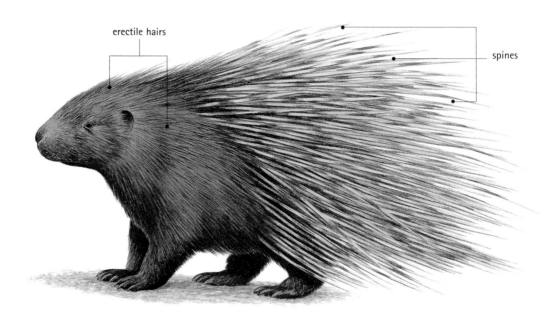

▶ **African porcupine**
Long black-and-white banded spines grow over the back and rump, and a crest of erectile hairs extends from the top of the head over the neck and shoulder region.

erectile hairs

spines

species of Indonesian porcupines are smaller and lighter but otherwise similar, and some zoologists group them in the same genus as crested porcupines.

The smaller brush-tailed porcupines have a more ratlike body form, with a relatively sleek body and a long tail with spiny tips. Brush-tailed porcupines are more agile and much faster, and are able to run and jump on their short but nimble legs. Brush-tailed porcupines, and the similar, but less obviously spiny, long-tailed porcupine regularly climb trees in search of food.

New World climbers

The North American porcupine has some of the ponderous, clumsy character of the crested porcupines, but like all the New World species it is adapted for life in the trees. It has strong, curved claws for gripping tree bark, and its

▲ *The North American porcupine is a careful but accomplished climber. It eats the inner bark of a variety of trees; as well as being nutritious, this bark helps keep the porcupine's teeth clean.*

IN FOCUS

Mossy camouflage

Like all porcupines, the hairy dwarf porcupine of the South American forests is mainly nocturnal and sleeps by day among tangled vegetation in the treetops. There, it is potentially vulnerable to airborne predators such as harpy eagles, which hunt by sight during the day. However, the hairy dwarf porcupine is camouflaged by the mixture of long hair and spines covering its body. The combination creates a mossy effect; as long as the animal stays still, it resembles a cushion of moss or lichen growing in the forest canopy, and it escapes detection.

◀ **South American tree porcupine**
Nine of the 11 species of New World porcupines—including this South American tree porcupine—have a prehensile (gripping) tail, which is used as an extra limb in the trees.

*The **first toe** is broad, increasing the width of the foot and therefore its grip.*

◀ **FOREPAW South America tree porcupine**
The bare palms and long, sharp claws of this species give it a firm grip when it is moving around in trees.

feet have naked soles with small fleshy ridges that enhance their grip on smooth, often wet trunks and branches of trees. The ridges enable the porcupine to climb slowly but surely into all kinds of trees to heights of 60 feet (18 m) or more. However, the North American porcupine is equally at home on the ground, especially in spring and summer, and this adaptability has allowed it to colonize a wide range of habitats throughout North America.

All the species of South American porcupines are suited to climbing trees. The aptly named prehensile-tailed porcupines and hairy dwarf porcupines (including the South American tree porcupine) have a long, muscular, mobile tail, which they use to grasp branches. In addition, these porcupines possess a modified first toe on each foot, which possesses a broad pad, increasing the width of the foot and therefore its grip. These highly arboreal species also benefit from having a lightweight build, often weighing less than 2 pounds (900 g). By comparison, an adult male North American porcupine can weigh up to 40 pounds (18 kg). Lightness gives the South American species much greater agility in the trees, although they move slowly in comparison with other arboreal rodents, such as squirrels.

Mountain mystery

The 11th New World species is the small stump-tailed porcupine, which lives in the mountain forests of central Colombia. It is a thickset animal resembling a large, spiny vole. As its name indicates, its hairy tail is short and only as long as its hind foot, certainly not prehensile like the tail of the other South American porcupines. Despite its lack of a prehensile tail, the stump-tailed porcupine is a good climber and lives mainly in the trees. Most aspects of its life are not well known, but its basic anatomy is similar to that of its South American relatives.

Skeletal system

CONNECTIONS

COMPARE the teeth of a porcupine with those of a *LION*. The two sets are completely different. Where the porcupine has enormous incisor teeth for gnawing, the lion has small ones; where the lion has huge canines for killing its prey, the porcupine has a gap; and where the porcupine has flat molar teeth for chewing, the lion has bladelike carnassial (cheek) teeth for slicing through meat.

Like other mammals, a porcupine has a strong bony skeleton. Each bone of the skeleton is formed from a combination of a hard but brittle mineral called calcium phosphate and a tough, flexible protein called collagen. Together they give the bones the rigidity needed to support the porcupine's body and a degree of springiness that stops the bones from breaking under the normal stresses of life.

Flexible chain

The main axis of the porcupine's skeleton is the arched spine: a chain of separate bones that could be compared to cotton spools threaded onto a length of string (the spinal cord). Each "spool" is called a vertebra, and the spools are linked together with tough, stretchy ligaments that allow the spine to flex. The North American porcupine makes good use of this flexibility when it climbs, grasping the tree trunks with its front limbs and arching its spine to draw the hind limbs upward. It then grips the bark with its hind feet and straightens its spine to push itself farther up the tree.

The North American porcupine's grip on the bark is enhanced by the long, strong, curved claws on each toe. It has four toes on each front foot and five on each hind foot. The other porcupines have the same arrangement, but in the tree-living porcupines of South America the first toe of each hind foot is relatively small and embedded in the footpad, so it is not very obvious on living animals. Like nearly all rodents—and indeed humans—porcupines are plantigrade: they walk on the soles of their broad feet rather than on their toes (digitigrade) like dogs and cats.

Gnawing teeth

The skulls of the larger Old World and New World porcupines are very different. Crested porcupines have a big, heavy skull, with thick

▼ **North American porcupine**
The North American porcupine has a compact skeleton with a long, flat skull; an arched spine; and stout limbs.

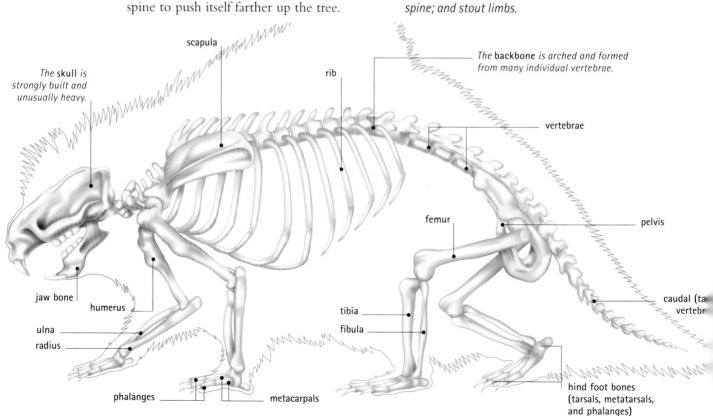

scapula

rib

The **backbone** *is arched and formed from many individual vertebrae.*

The **skull** *is strongly built and unusually heavy.*

vertebrae

femur

pelvis

jaw bone

humerus

tibia

caudal (ta... vertebr...

ulna

fibula

radius

phalanges   metacarpals

hind foot bones (tarsals, metatarsals, and phalanges)

bone and a domed cranium. By contrast, the North American porcupine has a smaller skull with lighter bone, a flatter cranium, and unusually broad cheekbones. The skulls could be thought to belong to completely different types of animals except for a feature that they share with all other porcupines—and indeed with all other types of rodents—their teeth.

A porcupine's skull is dominated by its long, curved, projecting front teeth: two at the upper tip of the snout and two at the bottom on the hinged jaw. These incisor teeth keep growing throughout the porcupine's life to compensate for the way they wear down as the animal gnaws its tough food. If the teeth do not meet properly and do not wear down quickly enough, they can grow out of control. If this happens, the teeth grow longer and longer, curving around until they kill the porcupine by locking its jaws together or even piercing its skull.

Porcupines also have 16 big cheek teeth—four on each side, top and bottom. They are ridged with hard enamel for grinding tough

Self-sharpening incisors

The gnawing action of a rodent like a porcupine keeps its incisor teeth sharp, as well as short. Each tooth has a layer of very hard enamel on the front but not on the back. So with use, the softer rear face of each tooth wears away faster than the harder front face. This creates an extremely sharp chisel tip at the front edge, which can hack through tough shoots, roots, and seeds. In the case of the beaver, these chisel-like teeth are sharp enough to cut down trees.

plant fiber to a digestible pulp. The cheek teeth are separated from the incisors by a gap called the diastema, and there are no conical canine teeth such as there are on many other mammals, particularly carnivores (meat eaters) like lions. The diastema makes gnawing easier; it also enables the porcupine to draw in its lips, closing its mouth behind the front teeth, if it is gnawing something inedible.

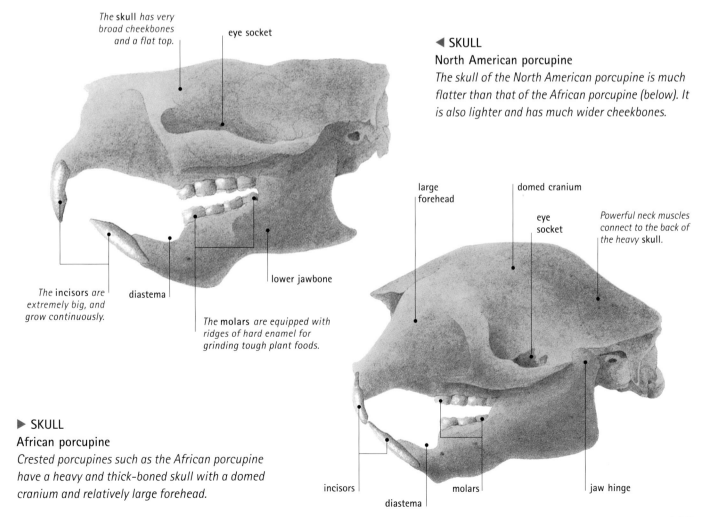

The skull has very broad cheekbones and a flat top.

eye socket

◀ SKULL
North American porcupine
The skull of the North American porcupine is much flatter than that of the African porcupine (below). It is also lighter and has much wider cheekbones.

The incisors are extremely big, and grow continuously.

diastema

lower jawbone

The molars are equipped with ridges of hard enamel for grinding tough plant foods.

large forehead

domed cranium

eye socket

Powerful neck muscles connect to the back of the heavy skull.

▶ SKULL
African porcupine
Crested porcupines such as the African porcupine have a heavy and thick-boned skull with a domed cranium and relatively large forehead.

incisors

diastema

molars

jaw hinge

Muscular system

In general, porcupines are slow-moving animals. They are vegetarians and therefore do not need to move fast to hunt prey; and, unlike many plant eaters such as gazelles and other antelopes, they do not depend on speed to escape their enemies. Instead, they rely on their spines to deter attackers. So porcupines are not particularly active animals, yet they still need highly developed muscle systems to operate their body.

A porcupine has two main types of muscles: smooth muscle and striated, or skeletal, muscle. Smooth muscle takes care of involuntary actions such as the rhythmic contractions that squeeze food through the animal's intestines. Skeletal muscles are those which are attached to its bones and which give it the power of movement. The other term for skeletal muscle, striated muscle, refers to its striped appearance when viewed under a microscope. The stripes consist of orderly bundles of long muscle fibers.

Muscle contraction

Each muscle fiber is made of alternating filaments of proteins called myosin and actin. When the fiber is activated by a nerve signal,

Prehensile tail

The tree-climbing porcupines of the New World genera *Coendou* and *Sphiggurus* have a long, prehensile tail that they use for support in the trees. The tail of the Brazilian porcupine, for example, is controlled by a series of strong muscles that act on the vertebrae of the tail, pulling them upward so the tail tip curls around a branch and grips it with a tough pad of skin on its upper surface. The muscles make up roughly half the weight of the tail, and, all together, the tail accounts for about a tenth of the weight of the entire animal.

projections from the thick myosin filaments attach to the thinner actin filaments and haul these thinner filaments alongside, so the actin filaments slide between the thicker myosin filaments. This action shortens the muscle fibers, making the muscle contract. When the

▶ REAR LIMB MUSCLES
North American porcupine
The muscular system of a porcupine's rear leg is typical of that of most mammals. Contraction of the biceps causes the leg to bend at the knee joint. The biceps work antagonistically against the triceps, contraction of which causes the leg to straighten.

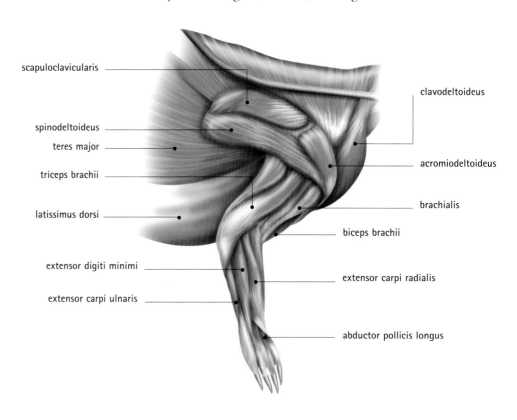

scapuloclavicularis — clavodeltoideus
spinodeltoideus —
teres major —
triceps brachii — acromiodeltoideus
latissimus dorsi —
brachialis
biceps brachii
extensor digiti minimi —
extensor carpi radialis
extensor carpi ulnaris —
abductor pollicis longus

Jaw muscles

Porcupines are classified among the cavylike rodents, which have a distinctive arrangement of jaw muscles. The main gnawing muscle on each side of the head is the deep masseter muscle, which is attached to the snout and passes through an aperture in the cheekbone before curving down to attach to the jaw. Squirrel-like rodents use a different muscle called the lateral masseter to do the same job; this muscle does not pass through the cheekbone. Mouselike rodents use both muscles for gnawing.

nerve signal is switched off, the links between the filaments are released, allowing them to slide apart again and extend. Smooth muscle also works in this way, although the filaments are not organized into orderly bundles.

Since muscles can generate power only by contracting, skeletal muscles are arranged in pairs that work in opposition. As one muscle of the pair contracts, it extends the other, which has relaxed. One member of the pair is often stronger than the other; the muscles that close a porcupine's jaws, for example, are much stronger than the muscles that open them.

Bristling spines

Most of a porcupine's muscles are typical of a wide variety of mammals, but certain muscles are unusually highly developed. Most South American porcupines possess very powerful muscles in their tail to help them climb.

All porcupines are equipped with a battery of muscles for bristling up their spines in self-defense. These are smooth muscle, which is stimulated automatically when the porcupine is alarmed—for example, by a predator. There is one muscle for each spine or spine cluster, and when it contracts it pulls on the spine root to lever the spine upright. In most other mammals—including humans—these muscles simply make the hair stand up, pushing up the hair root to create goose bumps.

▶ **QUILL ELEVATION**
When the arrector muscles are relaxed, the quills lie at an angle of 40 degrees to the porcupine's skin. When the arrectors contract, the quills are pulled into an upright position.

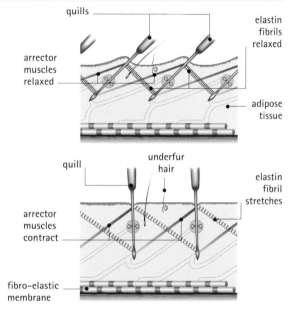

quills
elastin fibrils relaxed
arrector muscles relaxed
adipose tissue
quill
underfur hair
elastin fibril stretches
arrector muscles contract
fibro–elastic membrane

▶ *When threatened by a predator such as a lion, an African porcupine raises its spines to create a formidable defense. This is achieved by the contraction of smooth muscles at the base of the spine clusters.*

Nervous system

COMPARE the brain of a porcupine with that of a *RAT*. Despite their bumbling appearance, porcupines are actually quite intelligent and, like rats, they can be trained.

CONNECTIONS

A porcupine has a relatively large brain that controls a variety of body functions and conscious actions. The brain is the processing center of the animal's central nervous system (CNS), which also includes a bundle of nerve fibers that extends through its backbone as the spinal cord. The CNS, in turn, is linked to the peripheral nervous system (PNS); this system is a network of fibers extending to all the organs and muscles of the body.

The fibers consist of long nerve cells called neurons, which transmit electrical signals. Some of these neurons carry signals from sense organs and are known as sensory neurons. They pass information to interneurons in the CNS. The interneurons process that information, and trigger signals in motor neurons that, for example, make the muscles contract.

Coordination and control

Many of the functions under the control of the nervous system are automatic and largely unconscious, such as breathing and heartbeat,

Seductive scents

Like all rodents, porcupines rely heavily on scent for social interaction, especially courtship. Both males and females produce powerful chemical signals called pheromones, which not only attract the sexes to each other but also bring females into breeding condition, or estrus. So when a male is attracted to a female during the breeding season, her body automatically reacts to his strong scent signals by releasing an egg (ovulating), ensuring that pregnancy is a possibility after the mating.

and the reflexes, which, for example, raise the porcupine's spines when it is alarmed. These functions are controlled by interneurons in the brain stem and spinal cord. Coordinated body

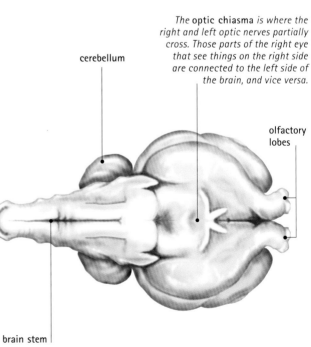

▼ BRAIN VIEWED FROM BELOW
North American porcupine

The **optic chiasma** *is where the right and left optic nerves partially cross. Those parts of the right eye that see things on the right side are connected to the left side of the brain, and vice versa.*

cerebellum

olfactory lobes

brain stem

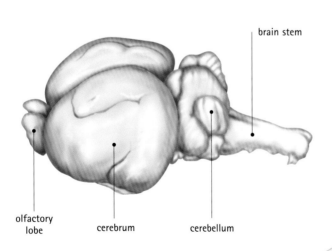

brain stem

olfactory lobe   cerebrum   cerebellum

▲ BRAIN VIEWED FROM THE SIDE
North American porcupine
The side view of the porcupine's brain shows its large cerebrum, which is involved in higher processes such as memory, learning, and interpreting sensory information. Of note is the large olfactory lobe, which interprets smell.

Nerve cells

Nerve cells, or neurons, are essentially long, tubular structures that extend into fine branches, or dendrites, at their ends. The dendrites at one end of the cell pick up electrochemical signals through contacts called synapses. The signal passes down the tube, or axon, and out through the dendrites at the other end. The neurons are often arranged in chains that pass the nerve signals from one neuron to the next. The axons of some neurons in porcupines, such as those of the spinal cord, are several inches long.

▼ Porcupines have small eyes and are nearsighted. They have keen hearing, but smell is their sharpest sense. At night, when porcupines are most active, they rely on their touch-sensitive whiskers to feel their way around.

movement is controlled by a part of the brain called the cerebellum; and voluntary, conscious actions are ultimately controlled by the largest part of the brain, the cerebrum.

The cerebrum also processes the data gathered by the porcupine's senses and stores some of the processed data in the form of memory. New World species such as the North American porcupine have been shown to have a good memory compared with other rodents and are able to learn simple tasks.

Porcupine senses

Porcupines gather sensory information using all the familiar senses of vision, hearing, scent, taste, and touch. Their vision is not particularly good, however, and several species including the North American porcupine appear to be very nearsighted. Like most mammals that are active mainly during the night, porcupines depend heavily on their touch-sensitive whiskers, their sense of hearing and, most important, their sense of smell.

North American porcupines constantly sniff the air for clues about their environment, food sources, the movements of other porcupines, and threats. Scent signals are an important means of communication, particularly between males. For example, old World crested porcupines mark their territories with pungent secretions from their anal glands, and males scent-mark more frequently than female porcupines. They memorize the scents of foods, enemies, and other porcupines, and the information gleaned in this way enables them to make choices and take appropriate actions.

Circulatory and respiratory systems

COMPARE the oxygen transport system of a porcupine with that of a *HAWKMOTH*. The moth's bloodstream does not carry oxygen. Instead, it has a system of tubes called tracheae that deliver air directly to the tissues.

The North American porcupine is no athlete. It moves slowly and clumsily and never has to chase prey or travel great distances, but it does climb tall trees, and if threatened by a predator such as a puma it may need to escape or drive the puma away with a swipe of its spiny tail. These activities use muscle power, and the muscles are fueled by blood sugar and oxygen in a process called respiration.

The sugar present in blood is called glucose, and it is a product of digestion. It is absorbed into the porcupine's bloodstream, along with the oxygen gathered by the lungs. Pumped by the animal's heart, the bloodstream delivers the glucose and oxygen to the porcupine's muscles, as well as to other organs such as the brain. When the sugar reaches the target cells, it is mixed with the oxygen to trigger a chemical reaction called oxidation, which breaks down the sugar into carbon dioxide and water. This process is called cellular respiration. It is essentially a slow, controlled form of burning, but instead of just releasing

energy in the form of heat and light, cellular respiration releases energy that can be used to power the porcupine's muscles and all its other body processes.

Arteries and veins

The oxygenated, sugar-rich blood is pumped to the muscles and other organs by the left-hand side of the porcupine's heart: a four-chamber structure made of muscle that has the virtually limitless stamina needed to keep working throughout the animal's life. The heart muscles exert pressure that pumps the blood through tubes called arteries, which have toughened walls.

The arteries deliver the blood to the tissues, where the oxygen and sugar are replaced by waste carbon dioxide and water. The blood then returns through veins to the right-hand side of the heart. On this return journey, the blood is under much lower pressure, and the heart action is supplemented by the effect of the animal's active muscles squeezing veins.

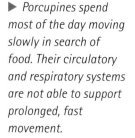

▶ Porcupines spend most of the day moving slowly in search of food. Their circulatory and respiratory systems are not able to support prolonged, fast movement.

The veins typically have relatively thin walls, and they contain valves that help keep the blood flowing in the right direction.

Air pump

The right-hand side of the heart pumps the deoxygenated blood into the lungs, where it exchanges its load of carbon dioxide for more oxygen. Each lung is a spongy mass of small, bubblelike, thin-walled air sacs called alveoli, which are surrounded by networks of fine blood capillaries. The porcupine's windpipe, or trachea, is linked to the alveoli by a branching network of air tubes of decreasing width called bronchi and bronchioles.

To breathe in, the porcupine contracts its muscular diaphragm, which is at the base of the sealed lung cavity. This contraction expands the spongy tissue of the lungs, so air is drawn in through the bronchial tubes. The inhaled air is relatively rich in oxygen, which passes through the thin walls of the alveoli and into the surrounding blood capillaries. Meanwhile, carbon dioxide and water pass out of the blood and into the inhaled air in the alveoli.

IN FOCUS

Air freshener

Although the porcupine draws in air with every breath, the inhaled air rarely reaches the alveoli where gas exchange takes place. The air in the alveoli barely moves at all, but if its oxygen has been used up, fresh oxygen diffuses into it from the inhaled air in the bronchial tubes. Similarly, if the air in the alveoli contains more carbon dioxide than the inhaled air, the carbon dioxide diffuses out into the bronchial tubes.

All these processes occur with every breath that the porcupine takes, and when it relaxes its diaphragm its lungs contract again, forcing the now deoxygenated, moist air out of its windpipe. The refreshed blood from the lungs returns to the left-hand side of the heart, which then pumps it to the muscles and other oxygen-hungry tissues.

▲ As with other mammals, porcupines are warm-blooded and maintain a constant body temperature. This characteristic, coupled with their thick coat of insulating fur, helps those living in colder climates survive through tough winters.

Digestive and excretory systems

COMPARE the digestive system of a porcupine with that of a ruminant such as a *GIRAFFE*. The giraffe's complex stomach allows it to digest leaves more efficiently than a porcupine, but it relies on the same kind of bacteria to break down cellulose.

All porcupines are basically vegetarian animals, feeding on a variety of plant material from tree bark, shoots, and leaves to fruit and nuts. Some species, such as the prehensile-tailed and hairy dwarf porcupines of South America, sometimes also eat insects and even small reptiles, and the Old World crested porcupines may do the same. Such animal prey, however, makes up only a very small fraction of their diet.

Tropical South American tree porcupines eat the same foods throughout the year, favoring the leaves, stems, flowers, and fruit that are always available in the tropical forests. Old World porcupines have a similar diet, although the ground-living crested porcupines also eat a lot of roots and bulbs—a diet that often brings them into conflict with farmers.

The North American porcupine inhabits the temperate and boreal zones, which are highly seasonal. The seasonal variation in food availability forces the porcupine to change its diet month by month. In spring, for example, there is a glut of tender young foliage, and a porcupine will even eat young grass in open meadows. In contrast, in late summer the foliage is tougher, but there are plenty of fruits, seeds, and nuts to be had. As fall gives way to winter, fruits and nuts become hard to find, and the porcupine switches to eating the evergreen needles of conifer trees, as well as gnawing tree bark. Porcupines can do a lot of damage to trees and even kill them by gnawing away all the bark within reach above the snow line.

Bacterial digestion

Foods such as bark and leaves—especially the needles of conifers—are tough and extremely difficult to digest. Their main ingredient is a fibrous carbohydrate called cellulose. This is a sugar compound similar to starch, but the sugar molecules are chemically bonded in such a way that the digestive enzymes of mammals cannot split them apart and turn them into simple sugars such as glucose.

▼ North American porcupine
As in other mammals, food is moved through the digestive tract by muscular contractions called peristalis. Since porcupines eat mostly plant matter they need a long digestive tract.

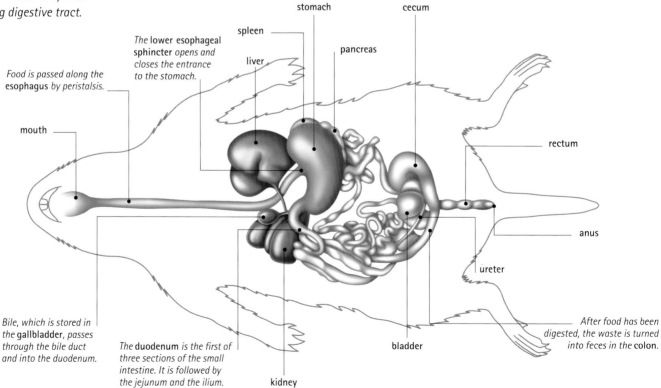

Food is passed along the esophagus by peristalsis.

The **lower esophageal sphincter** *opens and closes the entrance to the stomach.*

spleen

liver

stomach

pancreas

cecum

mouth

rectum

anus

ureter

Bile, which is stored in the **gallbladder**, *passes through the bile duct and into the duodenum.*

The **duodenum** *is the first of three sections of the small intestine. It is followed by the jejunum and the ilium.*

kidney

bladder

After food has been digested, the waste is turned into feces in the **colon**.

Some microorganisms do have this ability, including certain types of bacteria that enter the digestive tract with food and which can turn cellulose into sugar by fermentation. Porcupines—and many other plant-eating, or herbivorous, mammals—maintain populations of these bacteria inside their long digestive tract. The intestine of a North American porcupine is nearly 28 feet (8.5 m) long, and 46 percent of it contains cellulose-splitting bacteria. The porcupine digests the bacteria along with the sugar they release, but luckily the bacteria multiply fast enough to offset the rate of digestion, so the porcupine never eliminates its microbial digestive force.

Excretion of wastes

Food digested by the porcupine is absorbed through the wall of the intestine into its bloodstream. The blood then passes through its liver, which continues the work of turning the food into substances that the porcupine's body cells can use as fuel or to make tissues. The process creates waste products, including some poisons, which the cells of the liver turn into harmless substances. The bloodstream carries these metabolic wastes to the kidneys, where they are filtered out of the blood and mixed with water to form urine.

▲ *Porcupines that live in the rain forests of South America have a year-round supply of stems, leaves, and fruits to eat.*

IN FOCUS

A craving for salt

Animals need sodium, one of the elements of common salt, because it is essential for the functioning of their nerve fibers. Plants, however, do not need sodium, so many do not have any sodium in their tissues. As a result, animals that eat mostly plants (such as porcupines) may suffer from a lack of sodium. North American porcupines often develop a craving for salt and have been known to gnaw avidly at wooden tool handles to get at the salt left by the sweaty hands of workers. They also gnaw at bones and discarded deer antlers that they find on the ground to get extra supplies of other minerals such as calcium. However, some plants—such as yellow pond lily, aquatic liverworts, and arrowhead leaves—do have relatively high concentrations of sodium, and North American porcupines eat these when they are available.

Reproductive system

COMPARE the single, well-developed young of a North American porcupine with the litter of blind, naked young produced by a mouselike rodent such as the *RAT*. Many young rats do not survive, because they are so vulnerable to cold weather and predators.

Northorth American porcupines mate in fall or early winter, after the male has tracked down a female by following her scent. The two animals perform a ritual courtship, calling and dancing around each other. Before mating, the male typically showers the female with urine. This contains pheromones (chemical messages) that probably trigger ovulation, ensuring that there is a chance she will become pregnant.

Like all male mammals, a male porcupine is equipped with a penis that he uses to inject sperm into the female's reproductive tract. To achieve this he must mount the female very carefully and with her full cooperation, because he could easily be injured by her sharp spines. Even if she holds her spines flat it is still a risky procedure, especially for the long-spined Old World crested porcupines. So the female lies flat on the ground with her rump and tail raised, and the male mates with her while standing on his hind legs.

Soft spines

If the pheromones have done their work, the male's sperm encounter an egg or two released by the female's ovaries. A sperm and an egg fuse to create a fertilized cell called a zygote. If a zygote becomes implanted in the female's uterus it develops into an embryo that is linked to the uterine wall by a pluglike placenta and umbilical cord. This cord carries nutrients from the female's bloodstream into the developing embryo and also carries away waste products.

Like all cavylike rodents, porcupines have small litters of just one or two young after a long pregnancy. The African porcupine, for example, produces one or sometimes two young after about 16 weeks, and the North American porcupine has just one baby (rarely twins) after a very long pregnancy of 30 weeks.

Since they have so long to grow in the uterus, the young are very well developed when they are born, and even African

▼ MALE REPRODUCTIVE SYSTEM
North American porcupine
Male porcupines have testes, where sperm are made, and a penis to inject the sperm into a female porcupine's reproductive tract. The testes may be held externally (especially during the breeding season) within scrotal sacs, or internally within the abdominal cavity.

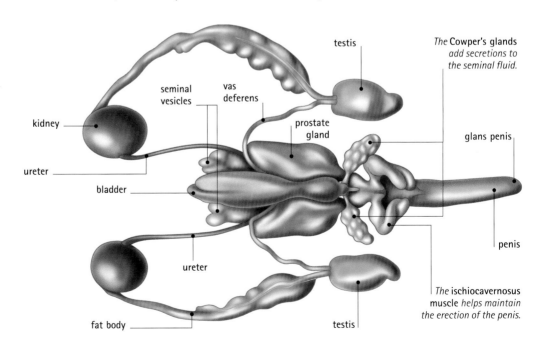

GENETICS

Strong stuff

The urine of strong, dominant male rodents has a more powerful scent than that of weaker males, indicating that it contains more of the pheromone that brings females into breeding condition. So females are more likely to ovulate and conceive when courted and mated by a strong, highly aromatic male. This increases the chances that their young will inherit the genes of strong males and survive to breeding age themselves. The effect of the pheromone is so powerful that exposure to a strong male's scent can make pregnant females of some species resorb (break down and absorb) their unborn young. They then ovulate again, so they are able to conceive and bear the young of the stronger male.

porcupines are born fully furred and able to see and walk. They also have spines, but luckily for the mother these are short and soft. The young are born headfirst, enclosed in a protective membrane. The spines start to harden a few hours after birth, by which time the young porcupine already has an instinct for presenting its spiny tail to any threat.

A newborn porcupine feeds exclusively on its mother's milk for the first few days; the nipples of a female crested porcupine are on the sides of her body rather than underneath, so she can suckle while lying on her stomach. Before long the young start eating solid food as well, and North American porcupines can even climb trees within a few days of birth. Along with their instinctive defense behavior, this gives them a higher than usual chance of surviving infancy, the most vulnerable stage of any animal's life.

JOHN WOODWARD

FURTHER READING AND RESEARCH

Hare, T. and M. Lambert (Eds). 1997. *The Encyclopedia of Mammals.* Marshall Cavendish: New York.

Kardong, Kenneth V. 1995. *Vertebrates.* William C. Brown Publishers: Dubuque, IA.

Macdonald, David. 2006. *The Encyclopedia of Mammals.* Facts On File: New York.

▲ *Porcupine young are born (after a long pregnancy) fully furred and able to see and walk. A young North American porcupine, as above, can even climb trees within days of birth.*

Potato

KINGDOM: Plantae FAMILY: Solanaceae
GENUS AND SPECIES: *Solanum tuberosum*

Potatoes are the world's largest food crop after rice and other cereals. The family of plants to which the potato belongs also contains many plants that produce well-known fruits and vegetables, including tomatoes, chilies, eggplants, and green and red peppers. Many plants in the potato family, such as tobacco and belladonna, produce potent chemicals that people use for medicinal and recreational drugs.

Anatomy and taxonomy

Botanists arrange species of plants into groups based on their similarities. However, they do not always agree on which characteristics should be used to judge similarity, so there are many systems in use that differ slightly from one another. Potatoes are classified in the family Solanaceae,

which includes many other useful food plants, such as tomatoes, peppers, and eggplants.

● **Plants** All true plants are multicellular (unlike most of the plantlike protists, which are single-celled). One of the key characteristics of most plants is that they are green. Unlike animals and fungi, most plants can make their own food by using the energy from sunlight to turn simple chemicals found in the air, soil, and water into carbohydrates (sugar compounds). This process is called photosynthesis (from *photo,* meaning "light"; and *synthesis,* "to manufacture"). These carbohydrates are the building blocks for all the materials that make up a plant. Chlorophyll, the compound that gives plants their green color, makes this possible by capturing the sun's energy.

● **Vascular plants** The simplest types of plants (nonvascular mosses and liverworts) do not have specialized tissues for transporting water and so are small and need to live in damp places. As plants evolved water-conducting tissues (a vascular system), they could grow larger and move into drier habitats. The first of these vascular land plants, or

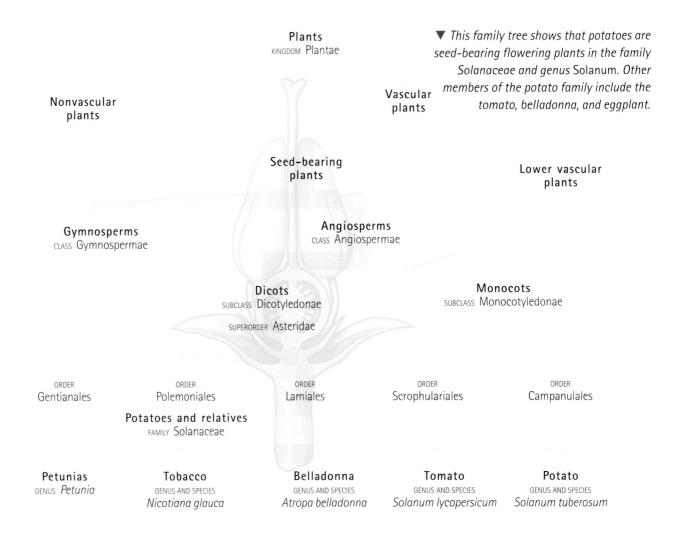

▼ *This family tree shows that potatoes are seed-bearing flowering plants in the family Solanaceae and genus Solanum. Other members of the potato family include the tomato, belladonna, and eggplant.*

Plants
KINGDOM Plantae

Nonvascular plants

Vascular plants

Seed-bearing plants

Lower vascular plants

Gymnosperms
CLASS Gymnospermae

Angiosperms
CLASS Angiospermae

Dicots
SUBCLASS Dicotyledonae
SUPERORDER Asteridae

Monocots
SUBCLASS Monocotyledonae

ORDER
Gentianales

ORDER
Polemoniales

ORDER
Lamiales

ORDER
Scrophulariales

ORDER
Campanulales

Potatoes and relatives
FAMILY Solanaceae

Petunias
GENUS *Petunia*

Tobacco
GENUS AND SPECIES
Nicotiana glauca

Belladonna
GENUS AND SPECIES
Atropa belladonna

Tomato
GENUS AND SPECIES
Solanum lycopersicum

Potato
GENUS AND SPECIES
Solanum tuberosum

tracheophytes, to evolve—around 420 million years ago—were spore-bearing plants (Pteridophyta), including ferns, club mosses, and horsetails. These plants reproduce using spores rather than seeds. Lower vascular plants with spores dominated the Earth until the seed-bearing plants evolved during the late Devonian period, between 375 and 360 million years ago.

● **Seed-bearing plants**   There are two major groups of seed-bearing plants: the gymnosperms (meaning "with naked seeds") that include conifers, such as pines; and the angiosperms (meaning "with enclosed seeds"). Angiosperms are flowering plants and have their ovules in an ovary that ripens into a fruit containing the seeds.

● **Flowering plants**   These plants make up the class Angiospermae, or angiosperms. They are loosely divided into monocotyledons (or monocots) and dicotyledons (or dicots). Monocots usually have long, narrow leaves with parallel veins. The flowers often have parts grouped in threes. Monocots include palms, grasses (including cereals), and orchids. Dicots are broad-leaved flowering plants. Their leaves have netlike veining and are highly variable in shape. The seedlings usually have two cotyledons (seed leaves). The flowers tend to have parts grouped in fours or fives. Familiar dicots include daisies, roses, and apple trees.

● **Potato family**   The potato family (Solanaceae) has about 3,000 species in 94 genera. They are widely distributed throughout tropical and temperate regions of the world. The highest diversity of species in this family occurs in South America, with 56 genera, 25 of which are found nowhere else in the world.

The potato family has edible, poisonous, and ornamental members, including tomatoes, eggplant, peppers, chilies, ground-cherry, tobacco, and belladonna. The plants grow as herbs, shrubs, and small trees. They are often spiny, and

▲ *There are hundreds of varieties of potatoes, including these long, knobby Rose Finn Apple fingerlings, which have yellow, waxy flesh and a creamy flavor.*

in many species the leaves are covered with fine hairs. The leaves can be simple, as in tobacco, or compound (divided into leaflets), as in the potato and tomato. The flowers are radially symmetrical (have more than one line of symmetry) and have five petals.

● **Genus *Solanum***   There are about 1,500 species in this genus. They grow on every continent except Antarctica, and are especially common in Australia and South America. Like other genera in the family Solanaceae, *Solanum* contains both poisonous and edible species. People eat the fruits of the eggplant and tomato, but the nightshades, such as belladonna, are toxic. There are also poisons in a potato plant.

● **Potato**   The potato is a major dietary staple in nearly all temperate countries. Almost 21 million tons (19 million tonnes) of potatoes were produced in the United States in 2005. Potatoes were domesticated at least 8,000 years ago in a region high in the Andes Mountains of South America. Through cultivation, potatoes spread through South and Central America and were introduced to Europe after the arrival of the Spanish. Potato breeding has produced hundreds of varieties. As well as the familiar brown-skinned and pale-fleshed potatoes, they can be reddish or even purple skinned; a few varieties have purple flesh. The potato is not closely related to the sweet potato, which is in the same family as morning glory (Convulvulaceae).

FEATURED SYSTEMS

EXTERNAL ANATOMY   Potato plants are herbs (plants that do not have woody stems) that grow up to 3 feet (1 m) tall. The potatoes are tubers (swollen stems) that grow at the end of thinner stems called rhizomes, underground. *See pages 918–922.*

INTERNAL ANATOMY   Like other dicots, potatoes have water- and nutrient-conducting vessels that run through the plant. In the stem, these are arranged in a ring, whereas in the root they form a core. *See pages 923–925.*

REPRODUCTIVE SYSTEM   Potato plants reproduce asexually by producing potatoes that survive over winter. They also reproduce sexually, producing a fruit that is similar to a small tomato. *See pages 926–927.*

External anatomy

CONNECTIONS

COMPARE the rhizomes of a potato with those of *MARSH GRASS*. Marsh grass rhizomes produce shoots from their tips, enabling the grass to spread very quickly. Potato rhizomes swell at the tip to produce tubers. These produce shoots only after winter when the parent plant has died.

COMPARE the veins of a potato leaf with those of an *ORCHID* leaf. Orchid leaves have parallel veins, whereas potato leaves have a network of veins branching from a central vein, or midrib.

Potatoes are perennial herbs: the plants can live for many years, dying down each winter. It is the swollen underground stem, or tuber, of the potato that lives through winter. In wild potatoes, the tuber overwinters underground, but in cultivated varieties, the tuber is usually dug up and stored. The tuber stores starch as an energy reserve for fast growth in spring. Potato plants grow to 1 to 3 feet (0.3 to 1 m) tall. They have straggling or relatively erect branching stems that never become woody. The flowers are about 1 inch (2.5 cm) across and come in many shades from white to purple. Below the ground, the plant consists of roots and also many underground stems called rhizomes. The potatoes form at the tips of these rhizomes.

Potato stems and leaves

Stems support the leaves and flowers and conduct water, nutrients, and products of photosynthesis between the roots and leaves. In most plants, stems are aerial, but potato stems may also grow underground. Stems grow in sections, with leaf joints, called nodes, separated by sections of stem called internodes. Buds form in the axils, the upper sides of the points where the leaf stems meet the plant stems. The buds grow into side branches.

Leaves are a plant's food factory, where photosynthesis takes place. This process converts chemicals from the air and soil into sugars using the energy of sunlight. Potato plants, like many other plants, tend to have thin, flat leaves that maximize the amount of sunlight they can capture. Potato leaves are divided into two to four pairs of leaflets, sometimes with smaller leaflets in between, and a single terminal leaflet. The leaflets are attached to a central leaf stem called a petiole. Unlike a leaf, a leaflet never has a bud in its axil.

Extensive root system

The roots anchor the plant firmly in the soil, forming an extensive branching network through which water and nutrients can be taken up into the plant. Potato roots are fibrous. Most potato plants grow from tubers and do not produce an initial taproot (main

▶ A "seed" potato is not actually a seed but a potato tuber that has survived winter to sprout new stems in spring. By the end of the growing season, a single "seed" potato may have produced— from its stems—up to 20 new potatoes.

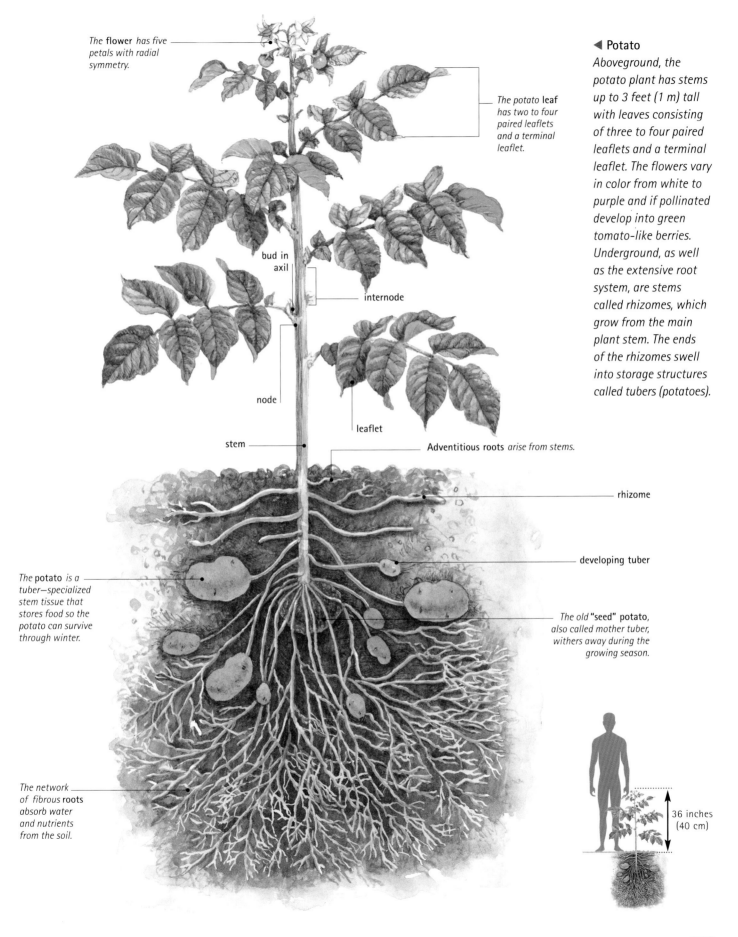

The **flower** has five petals with radial symmetry.

The potato **leaf** has two to four paired leaflets and a terminal leaflet.

bud in axil

internode

node

leaflet

stem

Adventitious roots *arise from stems.*

The **potato** *is a tuber—specialized stem tissue that stores food so the potato can survive through winter.*

The network of fibrous **roots** *absorb water and nutrients from the soil.*

rhizome

developing tuber

The old "seed" potato, also called mother tuber, withers away during the growing season.

◄ **Potato**
Aboveground, the potato plant has stems up to 3 feet (1 m) tall with leaves consisting of three to four paired leaflets and a terminal leaflet. The flowers vary in color from white to purple and if pollinated develop into green tomato-like berries. Underground, as well as the extensive root system, are stems called rhizomes, which grow from the main plant stem. The ends of the rhizomes swell into storage structures called tubers (potatoes).

36 inches (40 cm)

Going to seed

The buds on a buried potato remain dormant until spring. When the weather begins to warm, the buds begin to grow. A new shoot pushes up through the soil, and when it emerges into the light it produces leaves. As the aerial parts of the plant grow, producing side branches and more leaves, roots (called adventitious roots) grow out from the underground portion of the shoot's stem. The roots can emerge from anywhere on the internode section. Side branches also grow from the underground stem, from buds that form at nodes in the axils of leaf scales. These side branches form the rhizomes.

Young tubers start to form at the tips of the rhizomes, usually when the plants are 6 to 8 inches (15 to 20 cm) high, or after about five to seven weeks. The growth of the young tubers is the result of both cell division and expansion of cells. It depends on the plant's producing enough carbohydrate from photosynthesis to support the growth and metabolism of other tissues in the plant and having excess to store in the tubers. More than one plant can grow from each potato, as each eye (bud) can produce a new shoot. When the energy in the seed potato is expended and the tuber dies, there is no physical connection between these new shoots.

▲ After a dormant (inactive) period during winter, each "eye" (bud) of a potato develops into a new stem. In the growing season, this "seed" potato withers away, producing several clones of the parent plant.

root that develops when a seed germinates). The younger portions of the roots have root hairs. These are single-cell extensions of the root that together provide a huge surface area over which the plants absorb water and soil nutrients. Root hairs have a high turnover rate. After a short period they die and are replaced by new ones farther along the growing root.

Differences between roots and stems

Growing underground does not necessarily make a plant structure a root; stems can grow underground. And roots—for example, those of some species of orchids—can grow in the air. There are many differences between roots and stems. Stems have nodes, whereas roots do not. Stems have leaves or scales at the nodes, but roots have neither. Stems branch only in the axils. Roots branch much more irregularly, with branches arising from deep inside the root. The internal anatomy of roots and stems also shows characteristic differences in the arrangement of the vascular system (the water- and nutrient-conducting vessels). Roots have a central vascular cylinder—a core of tissues containing the vessels—whereas in stems the vascular tissue is in bundles either scattered (as in monocots) or in an intact ring, as in potatoes and other dicots.

Rhizomes: Underground stems

As well as the true roots, potato plants also produce underground stems called rhizomes, which grow as side branches from the buried portion of the main stem. Rhizomes have the external and internal structures of stems, not roots. Like aerial stems, they are divided by nodes. This is hard to see because they do not produce leaves. However, they do produce leaf scales, which are thin, brownish, and papery. These leaf scales do not persist for long. Often the only clue that they were present is a small curved leaf scar on the rhizome. Like aerial stems, rhizomes also branch at their nodes.

▶ POTATO TUBER

The potato tuber is the swollen end section of the underground stem, called a rhizome. The skin varies from brownish white to purple, depending on the variety. The potato's eyes are small buds that grow into shoots after winter.

Rhizomes do not have root hairs, so they do not contribute much, if anything, to the uptake of water and nutrients. However, by growing sideways into the soil, rhizomes help anchor the plant. The ends of the rhizomes swell to form tubers—called potatoes—which are the main storage site for carbohydrates and other nutrient reserves that the plant accumulates throughout its growing season.

Size and color of potatoes

Potatoes vary greatly in size, depending on the variety, the conditions under which they were grown, and the time of year of harvest. Some new potatoes are harvested when they

▶ POTATO BLOSSOM

Ranging from white through yellow to purple, potato flowers have five petals arranged around the reproductive parts (anthers and carpels). Insects, mainly bumblebees, pollinate potato flowers.

*A **lenticel** is a corky structure with pores that allow air into a potato tuber.*

*The **eyes** (buds) develop in a spiral pattern around the potato tuber.*

eyebrow (leaf scale)

skin

*Five **sepals** form the calyx.*

*In the center of the flower, the **anthers** (male reproductive parts) surround the carpel (female reproductive part) and are arranged in a cone-shape structure.*

pedicel

*Five **petals**, which are fused at their base, form the corolla.*

921

COMPARATIVE ANATOMY

Potatoes and sweet potatoes

Potatoes and sweet potatoes look similar and serve similar functions in their respective plants. Both are underground storage organs for carbohydrates and other nutrients. However, potatoes and sweet potatoes originate from different plant structures. A potato is a tuber (swollen underground stem), whereas a sweet potato is a swollen root. A potato has eyes (buds), whereas a sweet potato does not, because roots do not have buds on them.

► Sweet potato
The sweet potato is a swollen root—unlike the potato, which is a swollen stem. The potato and sweet potato are distant relatives; the sweet potato belongs to the family Convulvulaceae and the potato to the family Solanaceae.

▼ POTATO FRUIT
About 1.25 inches (3.2 cm) in diameter, the potato fruit is a green berry that resembles a small unripe tomato. Each potato berry contains up to 300 flat, oval seeds.

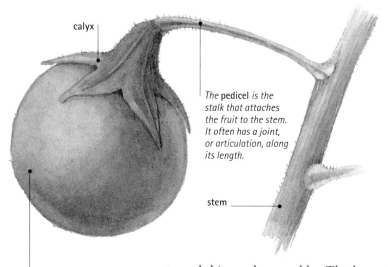

calyx

The pedicel is the stalk that attaches the fruit to the stem. It often has a joint, or articulation, along its length.

stem

The potato's fruit is a berry that contains seeds.

are not much bigger than marbles. The largest potatoes can reach an amazing 7 pounds (3.2 kilograms). A potato's skin color can vary from brownish white through pink to deep purple. The tissue inside the tuber normally ranges in color from white to yellow, but it, too, may be pink or even bluish.

The potato's eyes are small buds that grow into new shoots after the dormant (winter) period. Above each eye is an arching eyebrow mark. This line is the scar from the leaf scale. Sometimes, new potatoes still have membranous leaf scales attached, but these have usually been worn off by the time the potato reaches the store. Each eye always has an eyebrow because buds develop only in the leaf (or leaf scale) axil. Potatoes usually have many eyes, arranged in a spiral pattern that is clustered at one end (the tip of the potato), with an eyebrow (leaf scar) behind each eye. At the other end of the potato, there is a short section of rootlike stem (the rhizome), from which the potato grew, or a round scar where the rhizome was broken off.

Most potato tubers also have structures called lenticels on their surface. These contain pores through which air penetrates to the interior of the tuber. When potatoes have grown in wet soils with a restricted air supply, the lenticels are often large and corky.

Internal anatomy

The internal anatomy of the potato plant is based on a pattern shared with other nonwoody dicots. The aerial parts of the plant, especially the leaves, are photosynthetic. The stem of the plant has the mechanical function of supporting the leaves, so it needs to be able to withstand the wind and other physical forces. The stem also contains water-conducting vessels, nutrients, and the products of photosynthesis between the leaves, flowers, and roots. The roots do not carry out the same mechanical support function as the stems, so the internal arrangement of structures does not have to offer the same strength and flexibility. In the potato, underground stems, when swollen into tubers, perform a storage function.

Structure of leaves

Leaf tissue is composed of photosynthetic cells, a vascular system that transports water to these cells and takes away the products made by photosynthesis, and air spaces through which carbon dioxide and oxygen can diffuse.

A waxy cuticle on the leaf surface helps waterproof and protect the leaf. Underneath the cuticle, there is a single layer of cells called the epidermis. Dotted within the epidermis, mostly on the underside of the leaf, are pores called stomata (singular, stoma). Each stoma is surrounded by two guard cells. These kidney-shape cells change shape to open and close the stoma, controlling the plant's gas exchange.

CLOSE-UP

Photosynthetic cells

Photosynthetic cells are packed with chloroplasts. Chloroplasts are organelles— miniorgans inside cells that perform specific functions. The function of chloroplasts is to photosynthesize. Chloroplasts are packed with a green pigment called chlorophyll, molecules of which sit in membranes called grana. The pigment absorbs energy from the sun, and the chloroplast uses this energy to convert water and carbon dioxide into glucose. This simple sugar is then transported out of the cell to be used as an energy source or as a building block for parts in the plant.

COMPARE the tubers of a potato with the tuberoids of an *ORCHID*. Both structures are storage organs, but tubers form on underground stems, whereas tuberoids are modified roots.

CONNECTIONS

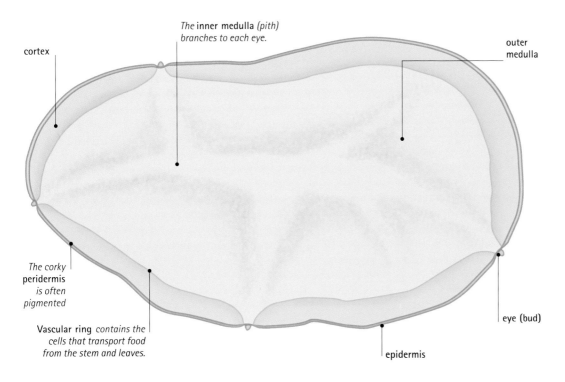

cortex

The **inner medulla** *(pith)* branches to each eye.

outer medulla

The corky **peridermis** *is often pigmented*

Vascular ring *contains the cells that transport food from the stem and leaves.*

epidermis

eye (bud)

◄ CROSS SECTION OF A POTATO TUBER
The skin has two layers: the epidermis and peridermis. Below these is the cortex, and below the cortex is the vascular ring. In the center of the potato tuber is the medulla. The cells of this tissue increase in size and number to store carbohydrate produced by the leaves. The inner medulla extends toward each eye, forming a continuous tissue that connects all the eyes.

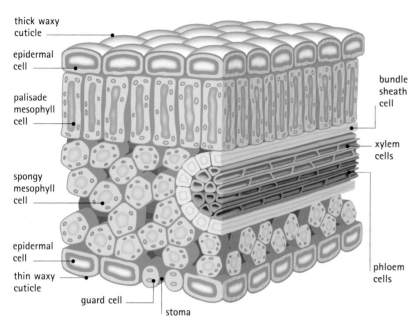

- thick waxy cuticle
- epidermal cell
- palisade mesophyll cell
- spongy mesophyll cell
- epidermal cell
- thin waxy cuticle
- guard cell
- stoma
- bundle sheath cell
- xylem cells
- phloem cells

▲ SECTION OF A POTATO LEAF

The leaf is made up of mesophyll tissue and veins, and encased by a layer of epidermal cells.

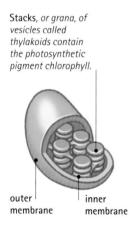

Stacks, or grana, of vesicles called thylakoids contain the photosynthetic pigment chlorophyll.

- outer membrane
- inner membrane

▲ ▶ CHLOROPLAST AND MESOPHYLL CELL

The mesophyll cells of leaves are packed with chloroplasts—organelles (miniorgans) that carry out photosynthesis.

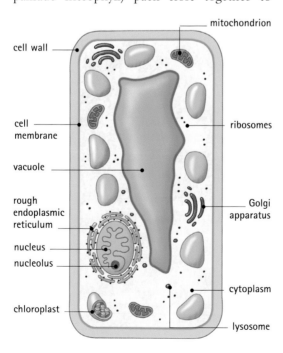

- cell wall
- cell membrane
- vacuole
- rough endoplasmic reticulum
- nucleus
- nucleolus
- chloroplast
- mitochondrion
- ribosomes
- Golgi apparatus
- cytoplasm
- lysosome

The vascular bundles in the leaves form the veins. The central vein, or midrib, has smaller secondary veins branching from it. In turn, smaller veins branch off the secondary veins, forming a network. As well as being a transport network, the veins also provide the leaf with some structural rigidity.

Most of the photosynthesis takes place in the middle layer of cells inside the leaf, called mesophyll. The cells of this tissue are packed with chloroplasts—the organelles in which photosynthesis takes place. Tall thin cells (called palisade mesophyll) pack close together to capture as much light as possible. Below these, there are spongy mesophyll cells, which are rounder and more loosely packed. Air spaces between these cells create channels from the stoma on the underside of the leaf though which air can diffuse.

Structure of stems

The layer of cells on the outside of the stem is the epidermis. As in leaves, the epidermis is covered with a protective waxy cuticle. Below the epidermis is the cortex. The outer cortex consists of a tissue called collenchyma. The thickened primary cell walls of collenchyma offer strength and protection. The inner cortex is a tissue called parenchyma; its cells do not have the same rigidity as outer-cortex cells. Cortical cells contain chloroplasts and so are green.

The vascular bundles of stems are arranged in a circle under the epidermis and cortex. They contain the conducting tissues of the stem. Phloem transports the products of photosynthesis throughout the plant. Phloem cells contain cytoplasm but lose their nucleus as the cells mature. These elongated cells have perforated ends called sieve plates where they connect to the next cell.

Water and minerals from the soil are carried in the xylem. The two main types of cells in the xylem are vessel members and tracheids. The vessel members, when mature, are dead empty cells, with highly lignified (woody) cell walls. These cells lose their end walls to form continuous tubes. Tracheids also conduct water and minerals, but these cells retain their contents, and so stay alive. They also retain their end walls; materials have to pass through pits at either end of the cells. As the plant grows, a tissue in the vascular bundle called cambium produces new secondary phloem and xylem. Phloem is normally deposited outside the cambium and xylem toward the center of the stem. The potato, though, is one of a small group of plants having both external and internal phloem associated with the vascular bundles. The center of the stem contains pith of large, unspecialized parenchyma cells.

Roots, tubers, and rhizomes

The root's internal anatomy differs from that of the stem mainly in the position of the vascular tissue. In roots, rather than forming a ring, the

◀ *The purple Peruvian potato has a dark purple skin and brighter purple flesh. Pigments called anthocyanins are responsible for the unusual color.*

vascular tissue forms a core through the center. This arrangement is easy to see in carrots. The xylem forms the center of the core, with clusters of phloem toward the outside. Encircling the conductive tissue is a layer of cells called the pericycle, which is the region where side roots are initiated (roots do not have buds). As they develop, side roots push their way through the cortex and epidermis. Tubers and rhizomes are modified stems, so they have the basic internal anatomy pattern of stems. But because they are underground, none of the tissues contain chloroplasts. Tubers are modified for storage and packed with swollen parenchyma cells.

The skin of a mature tuber consists of the epidermis and another band of several layers of corky cells below it called the peridermis. In most colored varieties of potato, the peridermis contains the pigment. Tissues from the vascular ring inward are called the medulla and form the fleshy part of the tuber. The medulla extends toward each eye, forming a continuous tissue that connects all the eyes of the tuber. The medulla is where the bulk of the storage takes place. Starch is stored as starch granules inside bloated parenchyma cells. Starch is made of glucose units linked together to form a stable, insoluble compound.

CLOSE-UP

Alkaloids: Poisons in the family Solanaceae

Members of the potato family produce potent chemicals called alkaloids. Most of these deter pests, but humanity has put them to its own uses. Many of these alkaloids have medicinal uses. Atropine from belladonna is used as a heart drug. *Solandra* species are a source of sacred hallucinogens in Mexico. Nicotine from *Nicotiana tabacum* is chewed or smoked as tobacco. It should be no surprise that the potato is capable of producing such toxins, given its close relationship to such poisonous plants as belladonna. All green parts of the potato plant, including tubers that have been exposed to light, contain the glycoalkaloid solanine. The green itself is just chlorophyll, but it does indicate that the plant is producing toxins, so people should avoid eating green potatoes.

▲ *The cherry-size berries of belladonna are often a cause of poisoning in children. The plant's leaves and roots (which are more poisonous) are a source of atropine—a drug used to treat various disorders, including heart problems.*

Reproductive system

CONNECTIONS

COMPARE the flower of a potato with that of the **APPLE TREE**. Both flowers have radial symmetry and five petals and sepals. The apple has an open, bowl-shape flower, with stamens free, whereas the potato has petals that are fused at the base and stamens that are fused into a cone-shape structure.

Potatoes reproduce asexually (vegetatively) and sexually. They reproduce asexually by using tubers—the edible part of the plant. Potato tubers are simply swollen extensions of normal vegetative tissue; no fusion of sexual cells is involved in their production. Potatoes are storage organs. While the rest of the plant dies in winter, the potato survives in the soil or dug up and stored and is able to grow again in spring. When potatoes are produced through vegetative growth, any plants that grow from them will be clones of the parent. Each plant can produce many potatoes, so the plants are a mechanism for multiplication as well as survival. However, unless they have been collected and stored by an animal, they do not usually move away from the site of the parent plant. Potatoes are a survival-through-time mechanism.

IN FOCUS

Potato varieties

Hundreds of varieties of potatoes have been cultivated over the years. The plants differ in their leaf shape, stem growth (tall or sprawling, for example), and flower color. The tubers themselves show a range of skin color and texture, flesh color, and shape. Some breeds of potatoes are more resistant to pests and diseases than others, and the degree of resistance to frost often dictates the regions in which different varieties of potatoes can be grown.

Potato plants are also capable of reproducing sexually. Sexual reproduction involves the mixing of genes (hereditary information) from the parents to produce new, genetically unique offspring. Sexual reproduction enables genetic diversity within a species. In potatoes, sexual reproduction involves the transfer of pollen

▼ SECTION THROUGH A POTATO BERRY
The fruit of the potato is a hard, green berry like a small tomato. It contains up to 300 seeds surrounded by a fleshy wall that develops from the ovary of the carpel.

stigma

anther

style

petal   filament

ovary

ovule

sepal

receptacle

pedicel

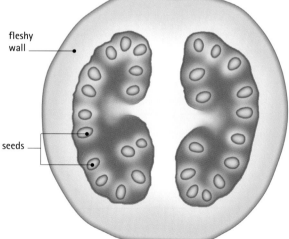

fleshy wall

seeds

▲ SECTION THROUGH A POTATO FLOWER
The male reproductive parts (the anthers) produce pollen and surround the female part (carpel). The carpel consists of a stigma (which receives pollen), style, and ovary (containing the ovules, which become seeds).

from one flower to the receptive female structures of another (pollination) by insects, most commonly bumblebees.

The color of potato flowers ranges from white through yellow to purple. Each flower has radial symmetry (more than one line of symmetry), and different parts of the flower occur in fives (five petals, sepals, and stamens). The sepals are the green structures that protect the bud. In ground-cherries, the sepals grow around the fruit to form a distinctive case. The five petals are fused at the base, and the yellow anthers (the male, pollen-bearing structures) are arranged in a closed ring to form a cone-shape structure. Hidden below this cone is the ovary (the base of the female structure). The stigma (the stalk that bears the receptive female surface) pushes through the cone of anthers, so it sticks out, ready to receive pollen.

▼ *The anthers form a cone-shaped structure surrounding the pollen-receiving stigma at the center of the potato flower.*

GENETICS

Polyploidy: A route to bigger vegetables

Chromosomes are threadlike structures present in cells with a nucleus, and they carry genetic information. Most animal and plant cells are diploid—they have two sets of chromosomes, a set inherited from each parent. However, some plants have more than two sets, and they are called polyploids. Diploid organisms normally produce haploid sex cells, or gametes—cells with only one set of chromosomes—in a process called meiosis. Errors in meiosis can lead to diploid gametes—with two sets of chromosomes, instead of just one. If diploid gametes fuse with normal haploid gametes, the resulting fertilized cells will have three times the haploid-chromosome number and are called triploids. Cells with four sets of chromosomes are called tetraploids. Many varieties of potatoes are tetraploids—their cells have 48 chromosomes, and they probably arose from ancestors with a haploid number of 12 chromosomes.

Polyploid plants are usually larger than their diploid parents, and often have larger fruits or heavier crops. For this reason, naturally occurring polyploids have long been selected by farmers. Now that the mechanisms are understood, polyploidy is exploited in modern breeding programs.

Sexual reproduction in potato plants results in many genetically variable seeds packed in a juicy dispersal package, the potato fruit. Potato fruits are berries that look like hard green tomatoes. Tomatoes, eggplants, and chilies are all fruits, too. Animals that feed on the fruits do not digest the seeds, which are later passed out in the feces. Thus these animals unknowingly act as agents for widespread dispersal of a new generation of potato plants.

ERICA BOWER

FURTHER READING AND RESEARCH

Bell, Adrian D. 1991. *Plant Form: an Illustrated Guide to Flowering Plant Morphology.* Oxford University Press: Oxford, UK.

Heiser, Charles B., Jr. 1987. *The Fascinating World of the Nightshades: Tobacco, Mandrake, Potato, Tomato, Pepper, and Eggplant.* Dover Publications: New York.

Heywood, V. H. 2006. *Flowering Plants of the World.* Firefly: Toronto.

Motley, T. J., N. Zerega, and H. Cross (eds.). 2005. *Darwin's Harvest: New Approaches to the Origins, Evolution, and Conservation of Crops.* Columbia University Press: New York.

Simpson, B. B., and M. Connor Ogorzaly. 1986. *Economic Botany: Plants in Our World.* McGraw-Hill: New York.

Puma

ORDER: Carnivora FAMILY: Felidae
GENUS AND SPECIES: *Felis concolor*

The puma is one of the most widespread mammals in the American continents. This large, powerful cat occurs from Alaska in North America to Patagonia in South America and has been recorded in habitats as diverse as forest, desert, and swampland, and from sea level to 14,500 feet (4,500 m). This huge range is reflected in its large number of common names. There are more names listed for the puma than any other North American mammal: cougar, mountain lion, California lion, panther, and catamount are all widely used, in addition to Native American and Spanish names.

Anatomy and taxonomy

Scientists categorize all organisms into taxonomic groups based on anatomical, biochemical, and genetic similarities and differences.

- **Animals** Pumas, like other animals, are multicellular and gain their food supplies by consuming other organisms. Animals differ from other multicellular life-forms in their ability to move from one place to another (in most cases, using muscles). They generally react rapidly to touch, light, and other stimuli.

- **Chordates** At some time in its life cycle, a chordate has a stiff, dorsal (back) supporting rod called the notochord that runs all or most of the length of the body.

- **Vertebrates** In vertebrates, the notochord develops into a backbone (spine) made up of units called vertebrae. The vertebrate muscular system that moves the head, trunk, and limbs consists primarily of muscles arranged in a mirror image on either side of the backbone. This arrangement is called bilateral symmetry about the skeletal axis.

- **Mammals** Mammals are warm-blooded vertebrates that have hair made of keratin. Females have mammary glands that produce milk to feed their young. In mammals, the lower jaw is a single bone, the dentary, hinged directly to the skull. This is a different arrangement from that found in other vertebrates. A mammal's inner ear contains three small bones (ear ossicles), two of which evolved from the jaw mechanism in the ancestors of modern mammals. Mammalian red blood cells, when mature, lack a nucleus; all other vertebrates have red blood cells that contain a nucleus.

- **Placental mammals** nourish their unborn young through a placenta, which is a temporary organ that forms in the mother's uterus during pregnancy.

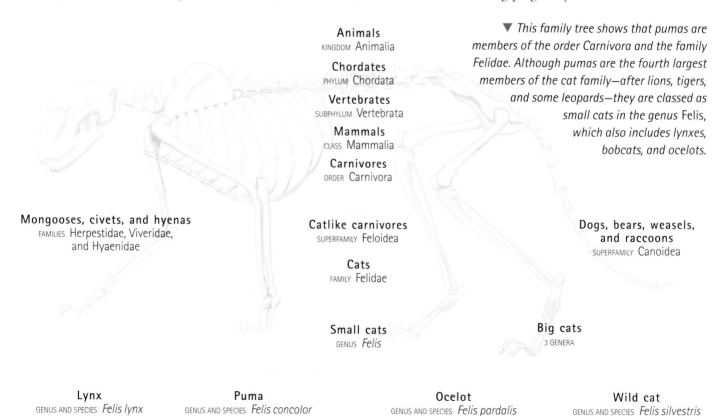

Animals
KINGDOM Animalia

Chordates
PHYLUM Chordata

Vertebrates
SUBPHYLUM Vertebrata

Mammals
CLASS Mammalia

Carnivores
ORDER Carnivora

▼ This family tree shows that pumas are members of the order Carnivora and the family Felidae. Although pumas are the fourth largest members of the cat family—after lions, tigers, and some leopards—they are classed as small cats in the genus Felis, which also includes lynxes, bobcats, and ocelots.

Mongooses, civets, and hyenas
FAMILIES Herpestidae, Viveridae, and Hyaenidae

Catlike carnivores
SUPERFAMILY Feloidea

Cats
FAMILY Felidae

Dogs, bears, weasels, and raccoons
SUPERFAMILY Canoidea

Small cats
GENUS Felis

Big cats
3 GENERA

Lynx
GENUS AND SPECIES *Felis lynx*

Puma
GENUS AND SPECIES *Felis concolor*

Ocelot
GENUS AND SPECIES *Felis pardalis*

Wild cat
GENUS AND SPECIES *Felis silvestris*

▶ *The puma is one of the most widespread mammals in the Americas, living in habitats as diverse as snowcapped mountains and tropical forests.*

● **Carnivores** The word *carnivore* can be used to describe an animal that eats meat, but it applies more specifically to members of the mammalian order Carnivora. Members of this group include dogs, cats, bears, hyenas, civets, mustelids, and raccoons. Most members of the group eat meat almost exclusively, but some, such as raccoons and most bears, have a mixed diet. A few carnivores, such as the giant panda, eat only plants. Carnivores have cheek teeth, called carnassials, that are specialized for cutting flesh. A characteristic of male carnivores is a penis bone, or baculum, that supports the penis and prolongs mating. In many species, it is this extended copulation that triggers the release of eggs from the ovaries.

● **Cats** The cats are the most committed meat eaters of all mammals. They have no teeth with chewing and grinding surfaces, so they cannot process plant food. A short muzzle, muscular jaws, and long, pointed canine teeth make formidable weapons for hunting and killing prey. All cats have a lithe, athletic body adapted for stalking and ambushing prey. Cats are fast, agile runners but lack stamina, so they tend not to chase prey over long distances. Most cats also climb well. There are four weight-bearing toes on each foot, each with a sharp, curved claw. Most cats have fully retractable claws (they can be drawn back into the toes); those of cheetahs are only partially retractable.

● **Small cats** Although similar in form to big cats such as lions and cheetahs, small cats are generally smaller and less powerful. They have a fully ossified (bony) hyoid bone in the throat, and so they cannot roar. They hunt small prey animals and usually kill with a swift bite to the neck. Small cats habitually adopt a crouching position when feeding. Their pupils contract in bright light to form vertical slits, whereas those of big cats remain round. All small cats currently belong to the single genus *Felis*, though some people believe they should be classified in several genera.

FEATURED SYSTEMS

EXTERNAL ANATOMY The puma is a large, powerful cat with a long body, long sweeping tail, and long back legs. The head is small and round, and the coat is typically golden or tawny with no obvious markings. *See pages 930–933.*

SKELETAL SYSTEM The skull is short, with a powerful jaw and teeth for killing and slicing meat. The spine is flexible, giving the cat great agility. *See pages 934–935.*

MUSCULAR SYSTEM One of the most powerful cats, the puma has especially muscular limbs. The hind legs are used for leaping and the front legs for wrestling large prey to the ground. *See pages 936–937.*

NERVOUS SYSTEM The brain and central nervous system process sensory information and signal appropriate responses. The brain is large, and the senses of hearing and sight are acute. *See pages 938–939.*

CIRCULATORY AND RESPIRATORY SYSTEMS The chest contains large lungs and a muscular heart. *See pages 940–941.*

DIGESTIVE AND EXCRETORY SYSTEMS Pumas eat exclusively meat, taken mainly from freshly killed prey. The gut is short. *See pages 942–943.*

REPRODUCTIVE SYSTEM Female pumas give birth to litters of cubs every other year. *See pages 944–945.*

External anatomy

▼ *The largest of the New World cats, the puma has a streamlined body with powerful limbs that enable bursts of speed when chasing prey. In addition, the long hind legs allow the cat to leap over obstacles as high as 6 feet (2 m).*

Most zoologists consider the puma a member of the genus *Felis*, the small cats, but unlike most other *Felis* species the puma is far from small. It is the fourth largest member of the cat family, exceeded only by the tiger, the lion, and some leopards. A large puma is considerably bigger than a jaguar and up to twice the size of an average snow leopard.

However, pumas vary greatly in size. Those in the far north and south of the species' range, such as Canada, Alaska, and Patagonia, can be up to twice as large as those in warmer regions, such as Ecuador and Florida. Growing big is a common adaptation for life in a cold climate. A large body retains heat more easily and can store reserves of fat and energy more

effectively. The puma is a powerful cat, but—as is true of all felines—its strength is tempered with grace. The body is long and narrow, with a deep chest and sturdy legs. The hind legs are larger and longer than the forelegs, reflecting the cat's impressive leaping ability. The tail is long and thick, and the black tip curves upward.

Feet and claws

The feet are broad and almost circular. The part of each foot that makes contact with the ground consists of a central pad and four toe pads. The first of the five toes on the front foot is modified into a dewclaw, a claw that does not make contact with the ground. The hind

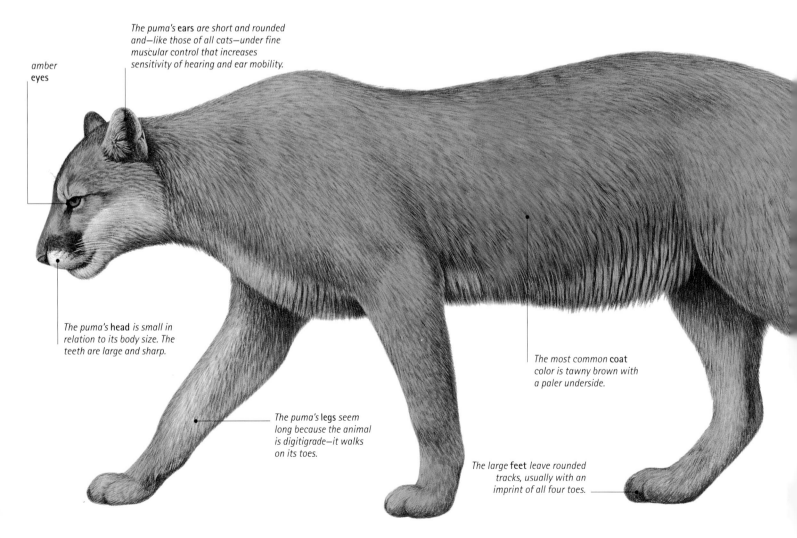

The puma's **ears** *are short and rounded and—like those of all cats—under fine muscular control that increases sensitivity of hearing and ear mobility.*

amber **eyes**

The puma's **head** *is small in relation to its body size. The teeth are large and sharp.*

The puma's **legs** *seem long because the animal is digitigrade—it walks on its toes.*

The most common **coat** *color is tawny brown with a paler underside.*

The large **feet** *leave rounded tracks, usually with an imprint of all four toes.*

feet have only four toes. The pads are formed from thickened skin and help cushion the cat's steps. The puma walks smoothly and quietly, even on hard surfaces, without the rattle or clatter often associated with the gait of other clawed animals such as dogs. The claws of a puma are curved and very sharp. They can be fully retracted into a sheath on the top of the toe, so that they do not snag during normal walking or inflict injury during nonaggressive interactions. Like other cats, pumas often leave deep scratch marks in wood and hard ground. This behavior is sometimes called sharpening the claws, but while it may help strengthen and clean the claws, its main function is territorial marking.

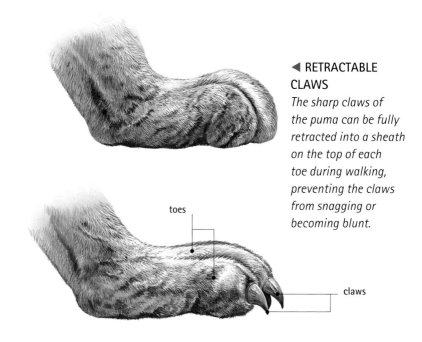

◀ RETRACTABLE CLAWS

The sharp claws of the puma can be fully retracted into a sheath on the top of each toe during walking, preventing the claws from snagging or becoming blunt.

toes

claws

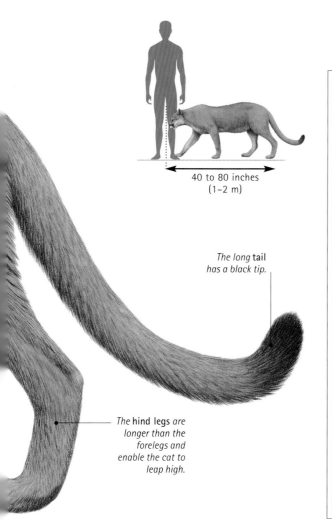

40 to 80 inches
(1–2 m)

*The long **tail** has a black tip.*

*The **hind legs** are longer than the forelegs and enable the cat to leap high.*

EVOLUTION

Spot the difference

Some biologists think that the ancestors of all modern cats were probably spotted, like a cheetah, a baby puma, or a young lion. Over many generations, the coat pattern was modified in different ways in different lineages: adult lions and pumas lost their spots; in leopards and jaguars, the spots opened up into rosettes; and in the ancestors of tigers and many small cats, the spots blended into stripes or tabby markings.

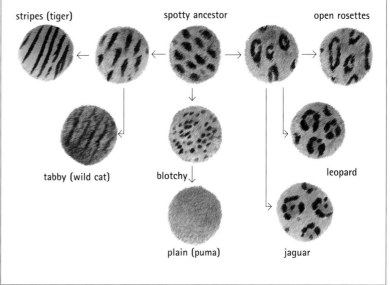

stripes (tiger)   spotty ancestor   open rosettes

tabby (wild cat)   blotchy   leopard

plain (puma)   jaguar

▲ BLACK FUR
The puma's coat color can vary. Black pumas do occur in the wild, but they are very rare.

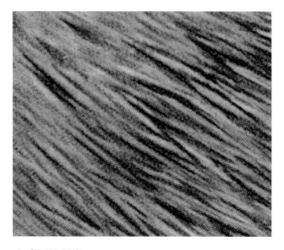

▲ GRAY FUR
In winter, the puma's coat may change in color from tawny brown to dull gray.

▲ REDDISH BROWN FUR
In summer, the puma's coat tends to deepen into a reddish brown or golden color.

COMPARATIVE ANATOMY

Big cat or small cat?

Zoologists have traditionally split cats into two main groups, the big cats and small cats. Big cats, mostly members of the genus *Panthera* (lion, tiger, leopard, and jaguar) are distinguished by: an elastic hyoid apparatus in the throat, giving them ability to roar but not purr; facial fur extending to the front edge of the nose; a tendency to eat lying down; and a pupil that contracts to a small circular point in bright light. The cheetah and clouded leopard do not roar but are generally considered to be big cats, too. Small cats, on the other hand, have the ability to purr but have a bony hyoid that prohibits roaring. Their facial fur stops short of the front edge of the nose, they eat in a sitting or crouching position, and they have a pupil that contracts into a vertical slit. On all criteria except size, the puma is allied with the small cats and is usually placed with them in the broad genus *Felis*, alongside the ocelot, lynxes, caracal, wildcats, and the domestic cat. However, some scientists place the species in its own genus, *Puma*.

The puma's coat is plain in adults, and usually a shade of golden or tawny brown. The color is darkest along the middle of the back and toward the tip of the tail. The ears are dark brown to black. The belly is pale beige to white. However, overall color may vary considerably between individuals and on the same animal at different times of year. In summer, the coat tends to be warm reddish brown or gold; it becomes much more gray and drab in winter. Interestingly, this is a range of seasonal coat color similar to that seen in many deer species, and it has much same implications for camouflage in a seasonally changing

GENETICS

Classification conundrums

Cat taxonomy is controversial. Few scientists believe all the small cats should be grouped in a single genus, but opinions vary greatly on how they should be divided. Thus *Felis* is usually used for them all by default. Scientists also disagree over the status of many cats: some think they are species and others think they are merely subspecies. The debate has important implications. For example, the Iriomote cat is a rare cat living on one Japanese island. Some people believe it is a subspecies of the leopard cat and so are not greatly concerned by its scarcity because the leopard cat is not threatened with extinction. However, if, as other scientists think, the Iriomote cat is a species in its own right, it is the world's rarest small cat and faces the imminent risk of extinction.

landscape. Gray blends well with bare trees and rocks, whereas golden or tawny shades provide better cover in sunshine-dappled summer grass. Young pumas are born with a spotted coat and dark rings along the tail. The markings last until the pumas are about six months old, when the juvenile coat is replaced with adult fur.

Skin care

The skin has many small sebaceous glands that seep greasy secretions for conditioning the coat. The puma spreads the secretions around with its tongue, which is covered with small, backward-pointing papillae. Thus, the tongue serves as a rough brush for grooming. The largest external glands are the anal glands, which are located in two muscular sacs that open under the tail; these sacs can be contracted suddenly to expel a foul smelling secretion that is used for marking a territory.

The cat's whiskers

Hairs are a unique mammalian feature, and the puma, like most carnivores, has several types of hairs. Two types of soft hair, called awns and down, contribute to a warm underfur. The underfur is covered by a sleek layer of longer, stronger guard hairs, which help keep the underfur clean and dry. It is mostly guard hairs you would feel if you were to stroke a puma's coat. There are also highly specialized hairs called tactile pili, vibrissae, or whiskers. Whiskers are thicker than normal guard hairs and more deeply embedded in the skin. The follicles of whiskers are richly supplied with blood and nerves and are sensitive to touch and pressure from movements of the air. Whiskers are found mostly on the puma's face, but they also occur elsewhere on the head and on parts of the legs and feet.

▼ *The puma's whiskers are sensitive to touch and pressure from movements of air. They help the animal feel its way through the undergrowth at night when hunting for prey.*

Skeletal system

CONNECTIONS

COMPARE the restricted movement in the shoulder and hip joints of the puma with the extensive rotation that can be performed by the same joint in a **FRUIT BAT** or a **CHIMPANZEE**.

Cats are known for their athleticism, much of which is owing to the flexible nature of the spine. Animals such as fish and reptiles have a spine that flexes only from side to side. But the spine of many athletic land-dwelling quadrupeds flexes not only from side to side but also up and down. The vertebrae of the puma's spine are rounded at the joints, allowing them to rotate smoothly against each other so the cat can take exceptionally long strides when running and leaping and also make spectacular midair twists and turns. This ability enables cats to land always on their feet after a leap.

Backbone and ribs

The puma's spine is the principal supporting structure of the body. It is made up of about 53 bones called vertebrae. Of these, there are seven in the neck, the first of which (the atlas) forms a hingelike connection with the skull that allows the head to nod up and down. The second cervical vertebra (the axis) allows the head to rotate on the neck. The cervical or neck vertebrae are followed by 13 thoracic vertebrae. These have long spinal processes to which the large neck and shoulder muscles supporting the head are attached. The thoracic vertebrae also bear ribs, which support and protect the deep chest cavity and lungs. The

first 12 ribs are also connected to a bony sternum, or breastbone, by short sections of cartilage. The 13th rib, known as a vertebral (or floating) rib, is shorter than the others and does not connect to the sternum.

Each thoracic vertebra has a pronounced dorsal spine, to which the powerful neck and shoulder muscles are attached. There are seven lumbar (lower back) vertebrae, with processes (points) sticking out to either side. The processes are anchor points for tendons connecting to major lateral and abdominal

▼ SKULL AND TEETH

The puma's skull is relatively short. The canine teeth are large for killing and the upper and lower carnassials shear through flesh like scissors.

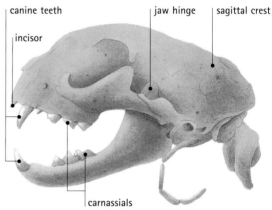

canine teeth | jaw hinge | sagittal crest
incisor
carnassials

▼ *Like all other species of cats, the puma has a very flexible spine. The puma's long limbs, with the elbow and knee joints close to the trunk, enable it to make long, fast strides.*

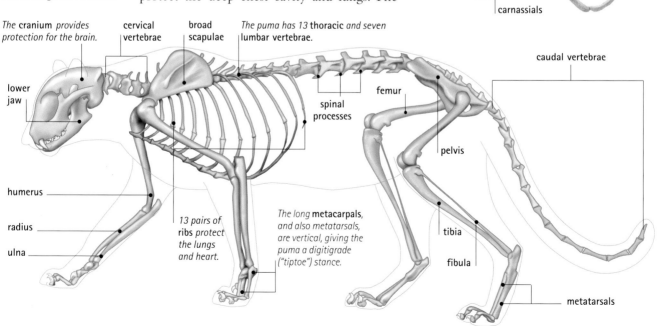

The **cranium** provides protection for the brain.

cervical vertebrae

broad scapulae

The puma has 13 **thoracic** and seven lumbar **vertebrae**.

caudal vertebrae

lower jaw

spinal processes

femur

humerus

radius

ulna

13 pairs of ribs protect the lungs and heart.

*The long **metacarpals**, and also metatarsals, are vertical, giving the puma a digitigrade ("tiptoe") stance.*

pelvis

tibia

fibula

metatarsals

muscles. Behind the lumbar vertebrae is a large single bone, the sacrum, formed from three individual vertebrae fused into one slightly flattened structure that articulates (hinges) with the other bones of the pelvic girdle. Finally, the puma usually has 22 or 23 caudal, or tail, vertebrae. The exact number of caudal vertebrae varies between and within species of small cats, but even species with a very short tail, such as the bobcat or lynx, have at least 14. Manx cats are a domestic breed in which the tail is reduced to a stump or is absent altogether.

Limbs and skull

As in all carnivores and many other fast-moving mammals, the puma's elbow and knee are located close to the trunk and articulate in a single plane—forward and backward. This type of movement enables the animal to make long, efficient strides. However, it cannot rotate its limbs as much as a human and cannot make extensive movements of the limbs out to the side. The price paid for speed is that without rotating shoulder joints, the movement of the cat's forelegs is restricted. However, the puma can rotate its lower forelimbs. The arrangement of the radius and ulna bones means that the forefeet can be rotated (making movements called pronation and supination). This ability allows the cat to clasp prey in front of it and make the swiping or swatting movements often used to knock down prey.

When pumas and other cats increase their stride length, only the fingers and toes touch

---

COMPARATIVE ANATOMY

The hyoid bone

Despite being similar in size to the Old World leopard, the puma differs in several important respects. Unlike big cats (in the genus *Panthera*) the puma cannot roar. The hyoid apparatus of pumas and other *Felis* species is fully ossified (turned to bone), so it lacks the flexibility needed to generate powerful reverberating roars.

▶ HYOID APPARATUS
The hyoid apparatus in the throat is made of bone in small cats. Because the bone is not flexible—unlike that in big cats—small cats cannot roar.

---

the ground. The long bones of the front paw and hind foot are held vertically, with the wrist and heel off the ground. This is called a digitigrade stance.

The puma's skull is short, with a large sagittal crest. Both these features strengthen the power of the animal's bite. The large crest allows for the attachment of huge masseter muscles, and the shortness of the jawbone maximizes the force these muscles can generate during biting. However, the short jaw allows room for only relatively few teeth. An adult puma, like other cats, has a total of 30 teeth, compared with 42 in a dog.

◀ *The puma's shoulder and hip joints move in only one plane—forward and backward—enabling the cat to make long strides. Over short distances, a puma can outrun prey such as deer.*

Muscular system

All cats are lithe, muscular animals. Their basic musculature is similar to that of other carnivores, with large blocks of striated muscles arranged symmetrically on either side of the body and concerned with voluntary movement and posture. Non-tiring smooth muscle, under involuntary control, regulates movements of the heart, gut, and other organs.

Some anatomists compare the axial skeleton of four-legged mammals to a girder bridge. However, this comparison hardly does justice to the power and flexibility of the puma's body. Combined with the major muscles of the trunk, a cat's body is more comparable to an archer's bow. The spine is the bow, which—owing to the ligaments and tendons connecting the vertebrae—is held in a more or less straight line. However, the two ends of the "bow" are pulled into a curve by large abdominal muscles, principally the rectus abdominalis, which connects the sternum to the pelvis. By putting the spine under tension, the abdominal muscles act like a bowstring, creating a large amount of potential energy—like that in a taut bow or a coiled spring. When the muscles are released, the spine flexes powerfully. In combination with similar mechanisms in the hind legs, this feature contributes greatly to the

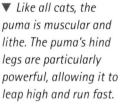

▼ *Like all cats, the puma is muscular and lithe. The puma's hind legs are particularly powerful, allowing it to leap high and run fast.*

IN FOCUS

Spring-loaded

Powerful hind legs give the puma exceptional leaping ability. From a standing start, a mature puma can leap into the branches of a tree 15 feet (5 m) above the ground. The long, powerful hind legs, sturdy front legs, and long, flexible spine also allow the puma to run very fast over flat ground, with a gait similar to that of a sprinting cheetah. Pumas tend not to chase prey for long periods and will flee if threatened themselves; pumas can outrun most pursuers.

strength and speed of movements such as leaping or pouncing.

Longissimus muscles

As in other cats, the longest muscle in a puma's body is the longissimus system, which runs along the back from the skull to the sacrum (where the backbone meets the pelvis) and beyond, into the tail. This extensive system helps control lateral (side-to-side) movements

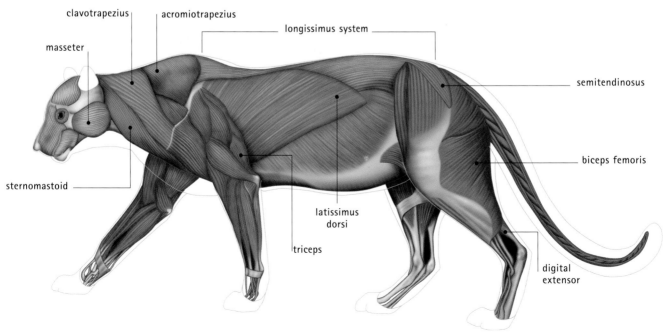

clavotrapezius · acromiotrapezius · longissimus system · masseter · semitendinosus · sternomastoid · biceps femoris · latissimus dorsi · triceps · digital extensor

Power and agility

Pumas are classified as small cats but are big and powerful enough to tackle large prey. If they are able to, pumas drag a large kill to a quiet place where it can be eaten, undisturbed, over several days. Shifting a dead weight such as a white-tailed deer is not easy, since the deer may weigh twice as much as the cat, so the puma needs to be strong. However, the puma's strength must be tempered with agility: if it were any bulkier it would struggle to leap, pounce, and climb with the necessary speed to catch any prey.

▼ *A puma catches a deer.*

of the backbone and resists the action of the large muscles beneath, helping the spine to spring back into shape and increase the animal's stride. The puma's longissimus muscles are also used to lift the tail.

The muscles of the cat's pectoral girdle, where the forelimbs attach to the body, are especially important because there is no bony articulation between the limb and the axial skeleton. Muscle and sinew, but not bone, hold the massive joints together. The clavicle, or collarbone, is present in cats but is not connected to the shoulder blade or sternum as it is in a human, for example.

Hair muscles

Erector pili muscles are tiny, but very numerous, with several thousands in a single square inch of skin. Each one is associated with a single hair follicle, and when the muscle contracts it makes the hair stand erect. In humans, the effect is called gooseflesh, or goose bumps. If you have ever seen a cat cornered by

a dog you will have witnessed a particularly effective example of erector pili in action. The cat stands tall, arching its back and often hissing or spitting. Almost all the hairs along the cat's back stand on end, making it seem much larger than it actually is. The behavior is an attempt to deter an attack.

▼ *The muscular, finely controlled limbs of the puma allow it to make big leaps from a standing start across rocks or into trees.*

Nervous system

Nerves are specialized "excitable" cells that are able to receive and transmit signals from one to another and stimulate activity in other cells, such as muscles or gland cells. The nervous system of higher animals is usually described in two or three parts. First there is the central nervous system (CNS), comprising the brain and spinal cord. The CNS comprises gray matter (nerve-cell bodies) and white matter. White matter is made up of the many complex processes (including dendrites and axons) that extend out of the cell body, making contact with other nerves and body cells.

All the other nerves connected to the CNS to and from other parts of the body are called the peripheral nervous system (PNS). A third component of the nervous system is called the autonomic nervous system. This describes nerves that are involved in involuntary control of the body—all the millions of signals that pass back and forth every moment to keep things functioning smoothly but without any conscious thought or effort. Antonomic nerves permeate every part of a vertebrate's body and form a large part of the CNS and PNS.

Structure of the brain

Cats are complex vertebrates with a highly evolved brain. The brain is the control center for the animal and it is supported in this vital role by a rich blood supply. The brain is also protected by a hard casing (the skull, or cranium) and layers of membranous connective tissue, the meninges. The first parts of the brain to evolve, many millions of years ago, were those at the base, close to the top of

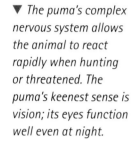

▼ *The puma's complex nervous system allows the animal to react rapidly when hunting or threatened. The puma's keenest sense is vision; its eyes function well even at night.*

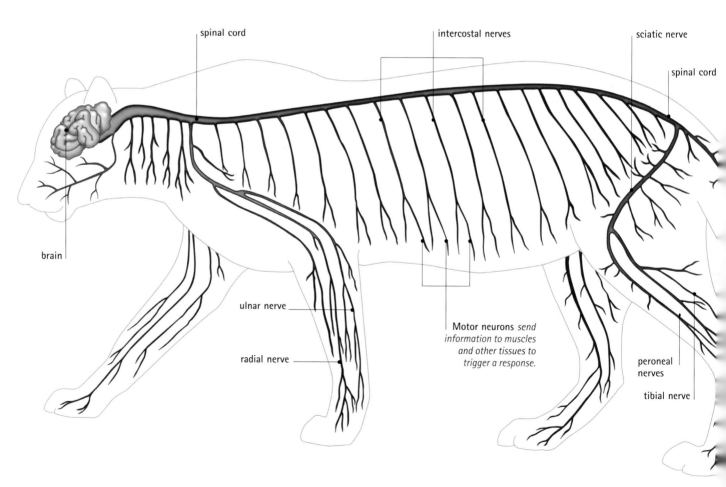

spinal cord

intercostal nerves

sciatic nerve

spinal cord

brain

ulnar nerve

radial nerve

Motor neurons *send information to muscles and other tissues to trigger a response.*

peroneal nerves

tibial nerve

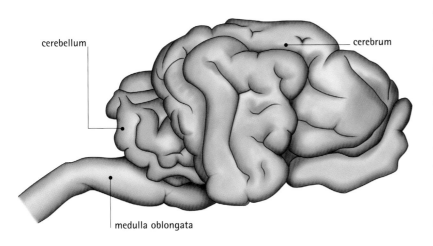

cerebellum

cerebrum

medulla oblongata

▲ BRAIN
Typical of mammals, the major part of the puma's brain is the cerebrum, which is concerned with higher functions, such as interpreting sensory information.

the spinal cord. The medulla oblongata (myelencephalon) is an extension of the spinal cord that is responsible for the most basic life-support systems, such as breathing, the beating of the heart, and swallowing. The next section of a vertebrate's brain is the hindbrain, or metencephalon, which controls basic movement (motor function). The midbrain, or mesencephalon, processes a limited amount of sensory information, but the senses are fine-tuned in the forebrain. The forebrain has two regions: the diencephalon, which serves as a vital relay center for sensory information and controls a large part of the endocrine system and autonomic nervous system; and the telencephalon, or cerebrum.

The two hemispheres of the cerebrum form the creased surface of the brain of higher mammals such as cats, and it is there that conscious perception and control of behavior

take place. The puma's senses of vision, hearing, touch, and smell are processed in the cerebrum. This part of the brain is also where memories are made and stored, skills are learned, and thoughts and intentions are translated into actions and behaviors.

Sharp sight
Vision is a puma's most important sense. Pumas are active by day and night, and their eyes are able to function well in a broad range of light levels. Perhaps surprisingly, a puma's sense of smell is poorly developed. Cats in general are less dependent on smell than other carnivores, in particular dogs. Within a cat's short snout there is little room for elaborate scent organs. Cats do, however, use scent to communicate: feces and urine are regularly used to mark territory.

IN FOCUS

Smart pupils

Usually, big cats have a pupil in their eyes that remains circular when it contracts (like that of humans), whereas that of almost all small cats contracts to narrow a vertical slit. The exception is the lynx, whose pupil is always round. Cats' eyes are able to see in a wide range of light levels, but especially at dawn and dusk. Muscles in the colorful iris allow the pupil to expand and contract, thereby controlling the amount of light that falls on the retina. The advantage of a slit pupil is that the aperture allowing light into the eye can be easily reduced to virtually nothing to protect the sensitive retina from bright sunshine. For the radially arranged muscles around a circular pupil, reducing the aperture to a pinprick would involve shortening the muscle fibers to almost zero length; such extreme shortening is difficult. In a slit pupil, the muscles are arranged more like curtains, which can be drawn across the pupil without excessive strain.

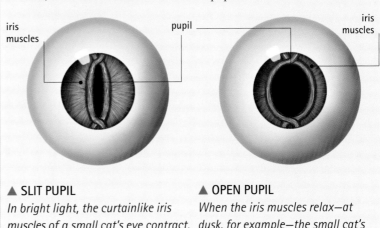

iris muscles

pupil

iris muscles

▲ SLIT PUPIL
In bright light, the curtainlike iris muscles of a small cat's eye contract, narrowing the pupil to a slit.

▲ OPEN PUPIL
When the iris muscles relax—at dusk, for example—the small cat's pupil expands, letting in more light.

Circulatory and respiratory systems

Since pumas are mammals, they are warm-blooded. Their blood contains red blood calls (erythrocytes) that lack a nucleus (unlike those of birds and other vertebrates, which are nucleated). The life span of a red blood cell without a nucleus is limited to about 120 days, so blood cells are continually renewed. Erythrocytes are red because they contain the respiratory pigment hemoglobin. Hemoglobin binds with oxygen when it is in an oxygen-rich environment (such as in lung tissue) and releases oxygen when it is in an oxygen-depleted environment (such as active muscle tissue).

Circulation of blood in the puma

Blood is pumped around the body by a muscular heart located in the center of the chest. The puma's heart is roughly pear-shape,

with the pointed end (containing the thick-walled muscular ventricles) directed backward and slightly to the left. Like other mammals, pumas have double circulation, with the right half of the heart receiving blood from the body and sending it to the lungs, and the left side receiving oxygenated blood from the lungs and pumping it vigorously around the rest of the body. Blood leaves the left ventricle at considerable pressure through a large artery, the aorta. The heart itself has its own blood supply: an offshoot of the aorta carries blood to the tireless heart muscles from where it drains back into the right-hand side of the heart ready to be pumped back to the lungs.

The puma's deep chest accommodates a large pair of lungs. When the puma needs to breathe in, muscles that are controlled by the autonomic

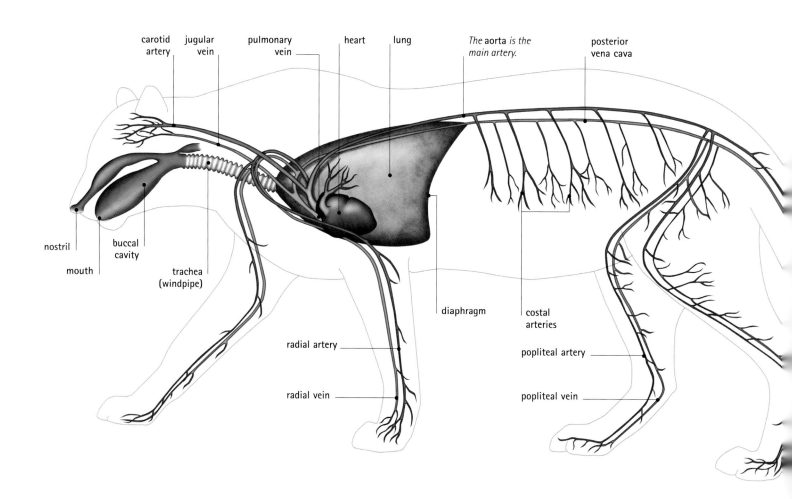

carotid artery • jugular vein • pulmonary vein • heart • lung • *The* aorta *is the main artery.* • posterior vena cava

nostril • mouth • buccal cavity • trachea (windpipe) • radial artery • radial vein • diaphragm • costal arteries • popliteal artery • popliteal vein

▲ *Pumas, like other mammals and birds, are warm-blooded. Because they create their own body heat, warm-blooded animals can survive in cold habitats, such as snowy mountains.*

(involuntary) nervous system act to expand the chest cavity. The rib cage is lifted upward and outward, and the diaphragm moves backward. The expansion of the lungs draws air in through the mouth and nose. The nasal cavity of the cat is kept open by a bone called the vomer. It is important that the nasal cavity remains open, since pumas kill large prey with a prolonged suffocating bite to the neck. Releasing the stranglehold to take a breath through the mouth is not an option—the prey may escape. However, the vomer is much smaller in cats than in dogs, reflecting the relative unimportance of the sense of smell to cats. Cats hunt by sight and hearing and have no need for a large nasal cavity lined with millions of scent cells.

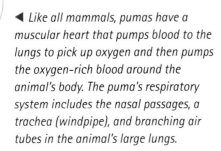

◀ *Like all mammals, pumas have a muscular heart that pumps blood to the lungs to pick up oxygen and then pumps the oxygen-rich blood around the animal's body. The puma's respiratory system includes the nasal passages, a trachea (windpipe), and branching air tubes in the animal's large lungs.*

IN FOCUS

Purr-fect sound

Although the structure of the hyoid apparatus in the throat does not allow small cats to roar, it does allow them to purr. Purring is a continuous low-frequency rumbling generated by vibrations of the hyoid on both the in and the out breath. Recently, scientists have begun to make links between purring and cats' famous ability to recover from serious falls and other injuries. Some researchers believe that ultrasound waves produced by purring help strengthen cats' bones and speed up the healing process. Physical therapists use ultrasound to treat injuries for the same reason.

Pumas generally do not make loud calls, although they do vocalize under certain circumstances. Loud eerie screams sounding something like *oooh-WAAhoo* are usually given under extreme stress—for example, when a puma is attacking or being attacked. Females in captivity have been heard responding to their young with mewing sounds like those of a domestic cat.

Digestive and excretory systems

COMPARE the bladelike carnassial teeth of the puma with the millstonelike carnassials of the **GRIZZLY BEAR**, used for grinding plant material.

COMPARE the short, simple intestines of the puma with the long coiled intestine of a herbivore, such as a **HIPPOPOTAMUS** or **ZEBRA**.

Pumas, and cat species in general, are almost exclusively carnivorous. A house cat, for example, needs at least 12 percent of its diet to provide protein, compared with 4 percent in the average dog. The only way to achieve this proportion of protein is to eat meat—and meat alone. However, pumas do not need to eat every day. Most individuals will eat several large meals in the days following the killing of large prey such as a deer but may not kill again for a week or two.

Pumas are not very selective about what they eat—almost any animal unlucky enough to get in the way is a potential meal. Deer are preferred because they provide food for several days, but other common prey animals include hares, agoutis, pacas, sheep, monkeys, llamas, young horses, hogs, peccaries, goats, skunks, rheas, raccoons, and wild turkeys. Puma feces often include the indigestible remains of feathers and even porcupine quills, proving that nothing of a suitable size is safe.

Teeth for hunting

The most important hunting tools at a puma's disposal are its teeth. The claws, although very sharp, are used more for climbing and fighting. Adult pumas have 30 teeth, including three pairs of incisors and one set each of canines and molars, in each jaw. Pumas also have three pairs of premolars in the upper jaw and two pairs in the lower jaw. Importantly, the fourth upper premolar and lower molars are slicing teeth called carnassials. These teeth are used to carve up the flesh after the prey is dead.

Pumas do not spend much time chewing. Their teeth are not well suited for masticating food. Instead, they rely on the potent mixture

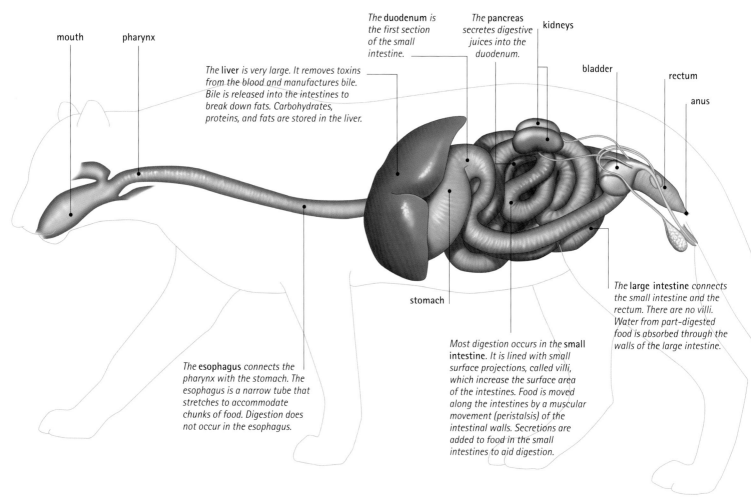

mouth   pharynx

The duodenum is the first section of the small intestine.

The pancreas secretes digestive juices into the duodenum.

kidneys

The liver is very large. It removes toxins from the blood and manufactures bile. Bile is released into the intestines to break down fats. Carbohydrates, proteins, and fats are stored in the liver.

bladder

rectum

anus

stomach

The large intestine connects the small intestine and the rectum. There are no villi. Water from part-digested food is absorbed through the walls of the large intestine.

The esophagus connects the pharynx with the stomach. The esophagus is a narrow tube that stretches to accommodate chunks of food. Digestion does not occur in the esophagus.

Most digestion occurs in the small intestine. It is lined with small surface projections, called villi, which increase the surface area of the intestines. Food is moved along the intestines by a muscular movement (peristalsis) of the intestinal walls. Secretions are added to food in the small intestines to aid digestion.

of enzymes and acid that is secreted by the stomach in order to break down the meat they consume. The entire lining of the stomach is glandular, and it begins producing digestive juices in anticipation of a meal. The cells of the stomach lining also secrete a layer of mucus, which helps prevent the digestive juices from attacking the puma's own tissues.

From the stomach, partially digested food passes into the duodenum. There it is blended with secretions from the pancreas and liver. Both secretions are highly alkaline, and they help neutralize acids from the stomach. Bile from the liver is stored in the gallbladder before being released into the digestive tract. In addition to producing bile, the liver also stores carbohydrate absorbed from the bloodstream and cleanses the blood of toxins. From the

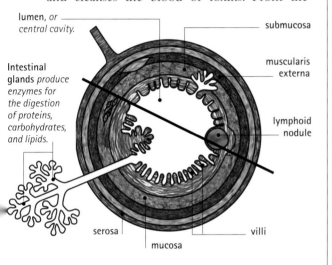

lumen, or central cavity.

submucosa

Intestinal glands produce enzymes for the digestion of proteins, carbohydrates, and lipids.

muscularis externa

lymphoid nodule

serosa

villi

mucosa

▲ CROSS SECTION OF INTESTINES
Concentric layers (mucosa, submucosa, muscularis externa, and serosa) surround the intestine. Many folds, or villi, substantially increase the surface area of the intestine over which digestive enzymes can operate on food. The section below the line depicts the main structures of the small intestine, and the section above represents the large intestine.

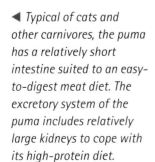

◄ *Typical of cats and other carnivores, the puma has a relatively short intestine suited to an easy-to-digest meat diet. The excretory system of the puma includes relatively large kidneys to cope with its high-protein diet.*

The killing bite

Pumas kill their prey in two ways. Large prey are brought to the ground and then suffocated with a viselike grip on the throat, and small animals are dispatched with a precise bite to the back of neck. The long, pointed canine teeth are well suited for the job: they slot into the gap between the victim's cervical (neck) vertebrae and force them apart. The spinal column is snapped if the vertebrae are forced apart by just a tiny fraction of an inch, and death is instantaneous. The roots of the puma's canines are well supplied with nerves, and it is possible that these help the puma feel the right spot to bite.

▼ *Pumas will kill and eat a wide range of animals, including deer, raccoons, and rabbits, which they kill with a sharp bite to the neck. They will even venture into fast-flowing rivers and streams to catch fish.*

duodenum, food passes into the small intestine, and the products of digestion (proteins, peptides and amino acids, sugars, fats, vitamins, and minerals) are absorbed into the bloodstream and transported around the body for immediate use or for storage. Waste continues through the large intestine and rectum and is expelled periodically through the anus as feces.

Metabolic and fluid waste is excreted by the kidneys. These are large in the puma and are located in the musculature of the lower back. A puma's kidneys receive about 25 percent of the blood pumped from the heart and filter out excess fluid and nitrogenous wastes. Urine manufactured in the kidneys is carried via the ureters to the bladder, ready to be passed out through the uretha.

Reproductive system

COMPARE the uterus of a female puma with that of a female *HUMAN*. A puma has a bicornate (two-horned) uterus, typical of mammals that give birth to several young at a time; whereas the human uterus is a single organ, suited to carrying one young at a time.

A male puma, like most small male cats, has a backward-pointing penis. This is an unusual feature among mammals, and in particular among carnivores, most of which have a forward-pointing penis. However, the position of the penis makes it possible for male cats to spray their pungent urine over vertical surfaces such as rocks and tree trunks without resorting to ungainly leg cocking or other gymnastics. By placing his mark high up on a tree or rock, a male puma ensures the best possible advertising position. Exposed to the breeze, a scent signal can travel many miles. A backward-pointing penis is of little use for copulation, so when the time comes to mate and the penile blood vessels engorge with blood, the penis swings around to the more common forward-pointing position. There is a small penis bone, or baculum, and the glans penis bears numerous horny papillae. The

GENETICS

Florida panther

Pumas were once widespread over most of North America, but since the arrival of European settlers their range has been pushed well back from much of the southeast. A tiny relict population has survived in Florida, where the species is called the Florida panther. However, Florida panthers are now perilously close to extinction. Part of the problem is that with such as small population, there is a severe risk of inbreeding. Conservationists have addressed this by importing pumas from Texas and have set up a puma sperm bank from which females can be artificially inseminated by unrelated males.

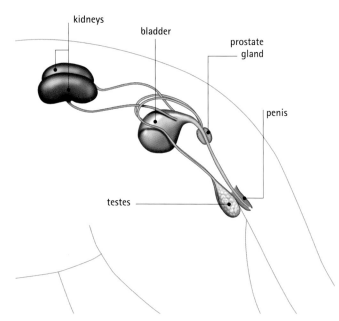

▲ MALE UROGENITAL SYSTEM
The male puma has a backward-pointing penis, used to spray urine on trees and rocks when it is marking its territory. However, the penis swings to a forward-pointing position before mating.

▼ FEMALE REPRODUCTIVE SYSTEM
Like many mammals, the female puma has a bicornuate (two-horned) uterus—an arrangement typical of mammals that give birth to more than one offspring.

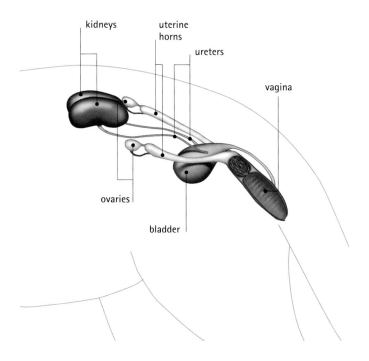

▲ *Female pumas give birth to litters of between two and six cubs. The cubs have spotted markings, which are lost at about six months of age.*

The mysterious onza

Mexican folklore tells of a mysterious big cat, the onza, which is said to look like a cross between a cheetah and a puma. Many people take this story seriously because there is fossil evidence for cheetahlike cats in the Americas throughout long periods of prehistory. However, one such animal was observed by expert zoologists, who concluded it was a young male puma that had fallen on hard times and had become very thin while in search of a breeding range of its own.

papillae probably help stimulate the female to ovulate (produce eggs) when they are scraped against the walls of the vagina as the male puma withdraws the penis.

The reproductive anatomy of a female puma is typical of most carnivores. The uterus is bicornuate—it has two long horns leading from the central chamber to the ovarian ducts. The bicornuate uterus is typical of species of animals in which several young are born in a litter. The horns expand to accommodate the growing embryos. The puma's vagina opens at the cervix ("neck") into a small chamber,

called the vestibule, which also accommodates the opening to the urinary tract. Female pumas have an estrous cycle lasting about 23 days, and a period of receptivity that lasts 4 to 12 (usually 7 or 8) days. They usually raise a litter every other year. Gestation lasts about three months, after which the cubs are born in litters of usually three or four (and up to six). Each puma cub weighs about 10 ounces (300 g) at birth.

Females reach sexual maturity at about two and a half years of age but do not breed until they have established themselves in a secure home range. Males usually breed first in their fourth year. Pumas live up to 20 years, though few wild individuals reach that age.

Like all mammals, female pumas suckle their young. Milk from mammary glands is drawn through tiny pores in the skin around the teats. Female pumas have three pairs of teats. The number of teats varies among *Felis* species. For example, the bobcat has only two pairs, and the wildcat has four pairs.

AMY-JANE BEER

FURTHER READING AND RESEARCH

Kitchener, A. 1991. *The Natural History of the Wild Cats*. Natural History of Mammals Series. Cornell University Press: Ithaca, NY.

Nowell, K., and P. Jackson (eds). 1996. *Wild Cats*. IUCN: Gland, Switzerland.

Sunquist, M., and F. Sunquist. 2002. *Wild Cats of the World*. University of Chicago Press: Chicago.

Rat

ORDER: Rodentia FAMILY: Muridae GENUS: *Rattus*

Rats are among the most successful of all rodents. The brown rat and its relatives the black rat and house mouse have adapted to live alongside people. By exploiting humans, rats have spread across the globe, even turning up in research bases on Antarctica.

Anatomy and taxonomy

Scientists categorize all organisms into taxonomic groups based largely on anatomical features (see below). Rats belong to a group of small rodents that also include mice, voles, and hamsters. Rodents are among the most numerous and successful of all mammals.

▼ *Rats belong to the subfamily Murinae, which contains about 530 species in 122 genera and constitutes the largest single mammal group. New members of this family are being discovered every year.*

● **Animals** All animals are multicellular. They get the energy and materials they need to survive by consuming the bodies of other organisms. Unlike plants, fungi, and the members of other kingdoms, animals are able to move around for at least one phase of their lives.

● **Chordates** At some time in its life cycle a chordate has a stiff dorsal (back) supporting rod called a notochord. Most, although not all, of the chordates are vertebrates. The notochord of vertebrates is surrounded by a spine. The spine is made from units called vertebrae (generally made of bone, though sometimes made of cartilage).

● **Mammals** One of the eight classes of vertebrates, mammals are warm-blooded animals with four limbs and, in most cases, a tail. They have body hair, which generally covers most of the body surface. All mammals nourish their newborn with milk secreted from glands on the female's front. A handful of mammals lay eggs, and another group, the marsupials, suckle their young in pouches on the underside of the body. Most mammals, including rats, nourish their unborn young through a temporary organ called the placenta, which allows the young to develop while still inside the mother.

Animals
KINGDOM Animalia

Chordates
PHYLUM Chordata

Vertebrates
SUBPHYLUM Vertebrata

Mammals
CLASS Mammalia

Rodents
ORDER Rodentia

Cavylike rodents
SUBORDER Hystricognathi

Mouselike rodents and squirrel-like rodents
SUBORDER Sciurognathi

Mouselike rodents
SUPERFAMILY Myomorpha

Pocket gophers
FAMILY Geomyidae

Rats, mice, lemmings, and hamsters
FAMILY Muridae

Jerboas
FAMILY Dipodidae

New World rats and mice
SUBFAMILY Sigmodontinae

Voles and lemmings
SUBFAMILY Arvicolinae

Old World rats and mice
SUBFAMILY Murinae

Hamsters
SUBFAMILY Cricetinae

Gerbils
SUBFAMILY Gerbillinae

● **Rodents** Most of the 2,000 species of this very large order of placental mammals are small. They are equipped with large chisel-like incisors, which make them expert gnawers. This ability has allowed rodents to exploit a huge variety of foods and adapt to life in a wide array of habitats.

● **Sciurognaths** This suborder Sciurognatha includes the squirrels, prairie dogs, beavers, and mouselike rodents, such as rats. Cavylike rodents are distinguished from the sciurognaths by having a larger head and more robust body. Cavylike rodents include guinea pigs (or cavies), chinchillas, capybaras, and porcupines. The sciurognaths are thought to resemble the primitive rodent form more closely.

● **Mouselike rodents** The superfamily Myomorpha alone comprises almost a quarter of all mammal species. Most of these species are rats and mice. The rest include dormice, jerboas, kangaroo rats, and gophers. Most mouselike rodents have only six cheek teeth on each side of the jaw—three above and three below. Squirrels and the other sciurognaths usually have eight. The jaw muscles of mouselike rodents also differ from other rodents. These muscles are attached farther forward on the skull than in other rodents, creating a very strong forward-thrusting biting action, which allows mouselike rodents to gnaw very effectively.

● **Muridae** This family includes the rats, mice, hamsters, gerbils, voles, and lemmings. Most of the group (1,000 out of 1,300 species) belong to the New World and Old World mouse and rat families, Sigmodontinae and Murinae respectively. Murids tend to be generalists, feeding on a wide range of foods when the opportunity arises. This diet is different from that of other myomorphids, such as dormice and gophers, which are adapted to more specialist lifestyles—living in trees and burrows respectively.

▲ *Brown rats (above) are rodents that are closely related to Old World mice. The brown rat and the house mouse are considered to be the most widespread land mammals.*

● **Murinae** This subfamily of rodents includes Old World mice and rats. Its original range covered Eurasia, Africa, and Australia, but several species, including the house mouse and brown rat, have spread throughout the world. The murines are successful because their basic body form allows them to adapt to almost any habitat. Jumping species, for example, have long limbs; burrowers are more robust. Murines also have a very short gestation period, which helps them reproduce rapidly and colonize new areas very quickly.

● **Rattus** This genus includes the brown rat and 55 other species of rats, including another globally widespread species, the black rat. Members of *Rattus*, like all rats in general, are larger than other murines and have pointed snouts.

FEATURED SYSTEMS

EXTERNAL ANATOMY Rats are large mouselike rodents. Coarse gray-brown fur covers the upperparts, and most species have paler fur on the underparts. Only the soles of the feet, the ears, and the tail are naked. *See pages 948–951.*

SKELETAL SYSTEM The skeleton of a rat is typical of a generalist rodent. Rats walk on the soles of their feet and have short legs and a long tail. These features make them good climbers and swimmers. *See pages 952–953.*

MUSCULAR SYSTEM The lower jaw muscles of rats and other mouselike rodents are attached to the front of the skull to produce a powerful forward-thrusting gnawing action. *See pages 954–955.*

NERVOUS SYSTEM Rats are clever animals. They can learn and remember complex routes and can navigate in total darkness, relying on their senses of smell and touch. *See pages 956–957.*

CIRCULATORY AND RESPIRATORY SYSTEMS Oxygen is absorbed into the rat's blood through two lungs. Blood is pumped around the body in vessels by the heart. *See pages 958–959.*

DIGESTIVE AND EXCRETORY SYSTEMS A rat's chisel-like teeth are used for gnawing on a wide range of food. Brown rats prefer meat but will also digest a wide range of vegetable matter. *See pages 960–961.*

REPRODUCTIVE SYSTEM Female rats can breed at a phenomenal rate: 18 hours after birth, the mother rat is preparing to mate. Each litter produces about eight pups. *See pages 962–963.*

External anatomy

COMPARE the tail of a rat with that of a *JACKSON'S CHAMELEON*. Both have a scaly-looking tail, but the chameleon's tail is also prehensile (able to grip onto objects). The rat's tail is not prehensile.

The brown rat is a large member of the mouse subfamily Murinae. The anatomy of a brown rat is typical of mice except that brown rats are about 10 times the weight of a house mouse. Male rats are up to 40 percent larger than females.

Brown rats originated in the forests of Asia, where a robust body was advantageous for rummaging through dense undergrowth. This and many other physical adaptations have also proved extremely useful in exploiting the habitats created by human populations. Although urban and agricultural habitats are very different from forests, for a rat the different environments represent a similar set of survival challenges. Cities, agricultural land, and forests are all complex environments with many places to search for food. A brown rat's tough yet agile body, combined with its keen senses and high intelligence, make it ideally suited for life in such challenging environments. Anywhere that people set up home, brown rats and other rodents are sure to move in, too.

The **fur** acts as insulation, both on land and in water.

The external portion of a rat's **ear** is large and prominent. Rats have a very good sense of hearing.

The **eyes** are farsighted and do not provide binocular vision.

Of all the rat's senses, smell is the most developed. The **nose** is the entrance to a long nasal cavity lined with cells sensitive to odors.

The **whiskers** are highly sensitive to vibration and touch.

The **hind paws** are much larger than the front paws and are used for locomotion and to paddle the rat forward in water.

The **front paws** are more dextrous than the hind paws and can be used to hold and manipulate food.

10 inches (27 cm)

Fur coat

Wild brown rats have dark gray-brown fur made up of coarse hairs. They may also have white or black blotches on their backs. The fur under the body is lighter—usually gray or tan. The hair's main function is insulation, keeping the animal warm, even when wet.

The brown rat's long association with people has led to the development of many domestic strains of brown rat. Some of these are kept as pets, but others are used in laboratory experiments. These domestic strains come in a wide variety of fur colors, from white to black.

Legs and paws

Rats walk on four legs. The forelegs are about two-thirds the length of the hind legs and considerably less powerful. On each foot, the five toes are tipped with clawlike nails. The forefeet are also used for holding food and nesting materials and as weapons (along with the teeth). The forefeet are also very flexible so the animal can twist and turn as it climbs.

Mouse forepaw

Rat forepaw

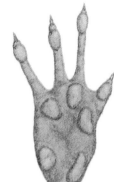

◀ PAWS

Mouse and rat

In mice and rats, the forepaws have only four digits, but the hind paws have five.

Mouse hind paw

Rat hind paw

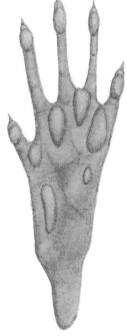

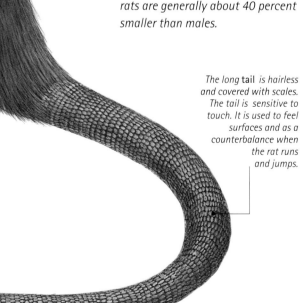

◀ Brown rat

Brown rats look much like very large mice. Most of their body is covered by course shaggy fur that is generally brown. The only areas not covered by fur are the nose, tail, and paws, all of which are sensitive to touch. Female brown rats are generally about 40 percent smaller than males.

The long tail is hairless and covered with scales. The tail is sensitive to touch. It is used to feel surfaces and as a counterbalance when the rat runs and jumps.

EVOLUTION

Rat ancestors

The first rodents are thought to have lived about 60 million to 55 million years ago. One of the best-known rodent ancestors is *Ischyromys*. This creature had a mouselike body with a large head and a long bushy tail, like a squirrel's. In common with the rodents of today, *Ischyromys* had large incisors useful for gnawing. Its hind legs were more powerful than its forelegs, suggesting that it was an active jumper. This trait, combined with its bushy tail, suggests that *Ischyromys* may have spent a lot of time up trees. Its short forelegs were probably used to grab food and reach for branches. Mouselike rodents, such as rats, are thought to have a single common ancestor. This ancestral rodent probably looked much like many modern-day mice. Almost certainly it was a very small animal, probably just a quarter of the size of a brown rat.

▶ INCISORS
Brown rat
The jaw opens on the underside of the rat's muzzle to reveal long chisel-shaped incisor teeth that are used to gnaw food.

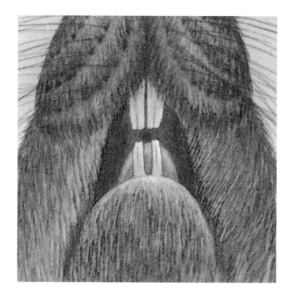

However, the feet of brown rats are not as flexible as those of other climbing rodents, such as squirrels, whose ankle joints can be rotated 180 degrees, allowing them to maintain a grip while running headfirst down tree trunks. A rat also uses its forepaws for digging.

The hind feet are at least twice the size of the forepaws. The hind feet are used primarily for locomotion. In water, they paddle the rat forward. On land, rats can rise up on their hind feet to investigate tall objects. The long hind feet also power short bursts of running and can produce a long jump. Many of the rat's relatives, such as the jumping mice, have evolved into specialist jumpers.

Tail

A brown rat's tail is not as long as the rest of the body. That is the most obvious difference between the brown and the black rat. These

▼ MUZZLES
Rat and mouse
A rat's muzzle is blunter and more wedge-shaped than a mouse's. The rat's ears are also comparatively smaller and sit closer to the head than a mouse's ears.

Rat

Mouse

Giant rats

One of the largest rat species in the world is the savanna giant pouched rat. These hefty rodents grow to about 17 inches (43 cm) from the tip of the nose to the base of the tail. The tail is generally about the same length again. Pouched rats are part of the murid family but belong to a different subfamily from most rats and mice. They live in forests and thickets in Africa south of the Sahara desert. Pouched rats feed on insects, seeds, and snails. The rats get their name from their large cheek pouches, which are used to store food. These pouches are similar to those found in hamsters, and some pouched rats resemble hamsters in other ways, too, such as having short tails. However, the savanna giant pouched rat is more ratlike, with a long tail, large ears, and a pointed snout.

two cosmopolitan rodents are often confused. As its name suggests, the black rat generally has darker fur, but that is not always the case. Both rodents are more or less the same size and found in the same habitats. However, a black rat's tail is always longer than the body.

Rats use their tail to help them balance. When swimming, brown rats hold the tail out of the water to counterbalance the rest of the body. The tail is also used as a counterbalance when the rat is moving through branches or in other precarious circumstances. The tail flicks from side to side to offset the body's weight. Along with the ears, and soles of the feet, the tail is largely naked. It is covered by coarse, grayish pink scaly skin. Without a covering of hair, the tail is more sensitive to touch, and the rat can use it as a feeler.

Mice and rats

The house mouse is anatomically typical of the subfamily Murinae. Rats are larger but have the same basic body form. However, there are a number of minor differences. For example, a rat's snout is blunter than the relatively pointed snout of a mouse. The rat's more robust snout allows the animal to rummage through leaf litter

COMPARATIVE ANATOMY

Jumping jerboas

Jerboas and kangaroo rats are often mixed up with each other. However, these mouselike rodents are grouped in two entirely different families. They are an excellent example of convergent evolution. This phenomenon occurs when two or more unrelated groups of animals evolve under similar conditions at different places around the world. Since their evolution is influenced by the same sort of environmental factors, the unrelated animals may end up looking very similar.

Jerboas and kangaroo rats both live in deserts; but jerboas are found in Africa and Asia whereas kangaroo rats live in the deserts of the southwestern United States and northern Mexico. Both types of rodent are nocturnal and have large eyes that can see at night. They have very large hind feet, short forelegs, and a long tail with a hairy tip. Both hop around on their hind feet like a kangaroo, using their forepaws to rummage about for seeds and other foods. The long tail acts as a counterbalance when the animal is jumping. It is also a useful prop when the rodent is standing on its hind legs. When threatened, jerboas and kangaroo rats bound off with powerful leaps. Some species can travel 10 feet (3 m) in a single jump—that is 15 times their own body length.

▼ *Like other kangaroo rats, Ord's kangaroo rat lives on a diet that consists mostly of seeds, from which it is able to extract the water and nutrients it needs to survive. Kangaroo rats have fur-lined pouches on the outside of their cheeks, within which they can store food and transport it to their burrow.*

or garbage more effectively. Also, a mouse's large ears have a prominent position on top of the head, ideally suited for detecting the slightest sound. In comparison with its head, a rat's ears are much smaller. They are also positioned lower down on the side of the head. Both rodents rely heavily on their sense of hearing, and a rat's ears are just as sensitive as a mouse's. It seems likely that although larger ears might be even more sensitive, they would be more easily damaged while the animal was moving through undergrowth or other small spaces.

Skeletal system

CONNECTIONS

COMPARE the gnawing teeth of a rat with those of a large carnivore such as a *LION* or *WOLF*. These animals have long canines for stabbing into flesh, and cheek teeth that slice food with a scissors action.

COMPARE the stance of a rat with that of a *ZEBRA* or *GIRAFFE*. The upright digitigrade (tiptoe) stance of zebras and giraffes is an adaptation for fast, efficient locomotion. In contrast, rats walk on the soles of their feet.

The skeleton of a brown rat is typical of rodents and small mammals in general. The body is close to the ground because the legs are short, and the animal adopts a crouching position when at rest. Since a rat is lightweight, contraction of the triceps muscles, which attach to the bones of the legs, results in rapid straightening of the folded legs so the animal can launch itself forward with the minimum of energy.

A rat's skeleton is made up of thin, lightweight bones, forming a strong but flexible internal framework for the body. The spine, for example, is able to bend considerably, making it possible for the rat to twist through winding, narrow holes. The tailbones are an extension of the spine, but they do not contain any of the spinal cord. The tailbones, too, form a very flexible unit.

The skull of a brown rat has several features that are typical of mouselike rodents. The lower jaw resembles that of a squirrel more closely than that of a cavylike rodent. Cavylike rodents have deep, robust jaws that form a blunt snout.

In mice, rats, and squirrels, however, the jawbones are thinner and form a more pointed snout. The zygomatic arches are curves of fused bones that extend along each side of the skull, forming the "cheekbones." In rats these are thick and form the widest part of the skull. This feature distinguishes mouselike and cavylike rodents from other rodents. The forward end of the arch flattens into the zygomatic plate. Between this plate and the eye socket is the infraorbital foramen, a hole in the skull through which blood vessels, nerves, and muscle fibers pass.

Plantigrade stance

Brown rats have four toes, or digits, on their forepaws and five on the hind paws. This arrangement is also seen in other members of

▼ **Brown rat**
The skeleton provides the stiff supporting framework against which the muscles can work. It also provides support for the body and protection to vital organs such as the brain and heart.

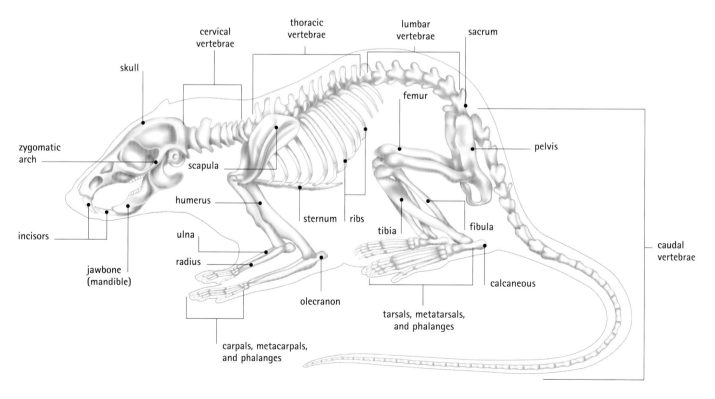

Brown rat

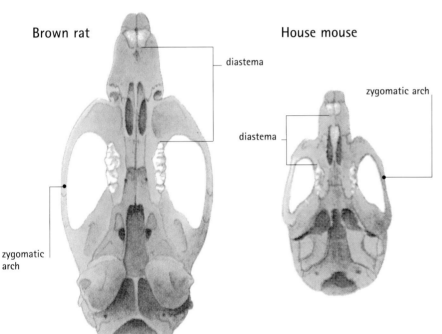

zygomatic plate
premaxilla
infraorbital foramen
frontal bone
parietal
maxilla
incisors
molars
zygomatic arch

Brown rat and house mouse
The skull of the brown rat and that of the house mouse are different in scale but very similar in form. Both show the pointed snout typical of mice and rats, and the large gap, or diastema, between the incisors and the molars. There are 41 bones in a brown rat's skull.

House mouse

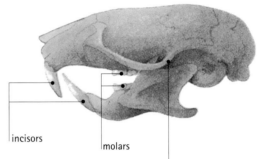

incisors
molars
zygomatic arch

Brown rat

diastema

zygomatic arch

House mouse

zygomatic arch

diastema

the subfamily Murinae. Rats are typical rodents because they have a plantigrade stance; that is, they walk on the soles of their feet. (Other plantigrade mammals include bears and humans.) A plantigrade stance produces a paw print that extends from the toes to the heel. Although the rat's stance prevents it from being a particularly fast runner, the flattened feet provide a wide, stable platform. The comparatively large surface area of the feet is useful for moving across unstable ground or climbing along a branch or ledge.

▲ SKULLS FROM BELOW
Brown rat and house mouse
With the lower jaw removed, the arrangement of incisors and cheek teeth is clearly visible There are two incisors on the upper jaw and three pairs of cheek teeth.

CLOSE-UP

Dentition

Like all rodents, rats have just four incisors. The teeth have a covering of hard enamel on the outer surface, but the inner surface encloses a relatively soft material called dentine. The dentine is constantly worn away as the animal bites the teeth together, producing a razor-sharp cutting surface of enamel. The teeth continue to grow throughout life to compensate for this wear. The word *rodent* is derived from *rodere,* Latin for "to gnaw." The gnawing action is enhanced by the fact that there is a large gap, or diastema, between the incisors and the rest of the animal's teeth. This gap is located where other mammals would have other incisors, canines, and premolar teeth. The gap helps the rodents gnaw on things from a range of angles. Behind the diastema, rats have six cheek teeth in each jaw, three on each side. The cheek teeth each have three ridges, or cusps, arranged in rows. These cusps crush the food as the upper and lower teeth meet when the jaw is closed. Owing to the structure of the jaw, the cheek teeth cannot grind food with a sideways movement of the jaw.

Muscular system

COMPARE the way a rat's muscles pull on a rigid bony skeleton with the way the muscles of an *EARTHWORM* or *STARFISH* squeeze against fluid-filled spaces.

COMPARE the gnawing jaw of a rat with the mouthparts of a *HAGFISH*. This fish is one of the few vertebrates that do not have a hinged jaw. Instead, it has a circular mouth, which is lined with horny teeth.

Rats, like all mammals, have three types of muscles. Cardiac muscle is found only in the heart. It is different from other muscles because it never completely rests during an animal's lifetime. The second type of muscle is smooth muscle. This muscle lines the blood vessels, exocrine gland ducts, and gut and is used to push food down the throat, through the stomach and intestines, and then out of the body again. Smooth muscle is also called involuntary muscle because animals do not have direct control over its contractions. The third group is skeletal muscle. This is involved in locomotion and other body movements.

Moving the body

Skeletal muscles work in pairs, each pulling in the opposite direction from the other. Each muscle is attached to a point on the skeleton, which acts as a rigid structure for the muscles to pull against. Bones are joined to muscles by tendons. Tendons are inelastic fibers made of the protein collagen. When a skeletal muscle contracts, it pulls on the tendon and causes the bone to move in one particular direction. If that muscle relaxes and its opposing partner contracts, the bone will move in the other direction.

Muscles are a mass of fibers made from two types of protein that lie in long filaments side by side. When the muscle receives a signal from the nervous system, the two filaments ratchet past each other, making the overall fiber shorter. On a larger scale this makes the muscle contract and exerts a considerable force.

Jaw muscles

Most of a rat's skeletal muscle system is very typical of a small mammal. However, the arrangement of the muscles in the lower jaw is unique to rats and mice. The upper jaw of all

▼ **Brown rat**
Apart from the unusual masseter muscles of the jaws, the rat's muscular system is similar to that of other small mammals.

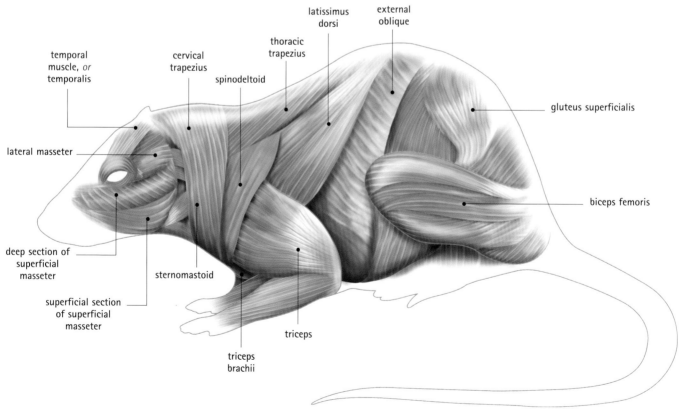

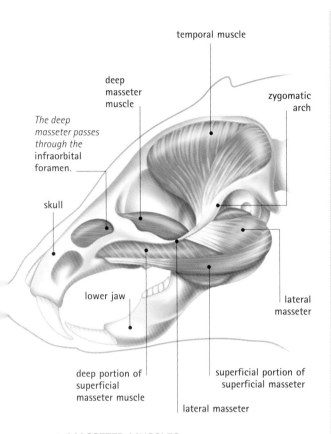

temporal muscle

deep masseter muscle

zygomatic arch

The deep masseter passes through the infraorbital foramen.

skull

lower jaw

lateral masseter

deep portion of superficial masseter muscle

superficial portion of superficial masseter

lateral masseter

▲ MASSETER MUSCLES

The brown rat's strong masseter muscles connect the lower jaw to the skull. The masseters provide the force for the powerful gnawing motion the rat makes when eating.

mammals is fused to the skull and cannot be moved. Thus all the biting power has to be generated by movement of the lower jaw. The muscles that move the lower jaw up and down are the deep, lateral, and superficial masseters. The masseters connect the lower jaw to the skull. In mouselike rodents these muscles are connected much farther forward than in other animals. This has the effect of producing a powerful up and forward biting action. With each bite, the lower incisors scrape against the underside of the upper ones, creating the gnawing action characteristic of rats.

The superficial masseter connects the middle of the lower jaw to the maxilla—an area near the front of the skull. The lateral masseter is connected to the zygomatic arch and plate. The deep masseter passes behind the zygomatic arch and through the infraorbital foramen before connecting to bone near the top of the skull.

Tail tales

A brown rat's tail is highly flexible. It is used to counterbalance the body and as a sensitive feeler. However, different rodents use their tails for a variety of other tasks. For example, the European harvest mouse has a prehensile (grasping) tail that it curls around stalks. This "fifth limb" helps the mouse clamber through flimsy vegetation. Another specialized rodent tail belongs to the beaver. This is broad and flat and is used as a powerful paddle. The tail of another aquatic rodent, the muskrat, has flat sides. The muskrat uses its tail as a rudder to steer itself through water. Hopping rodents, such as jerboas, have a long tail tipped with a clump of hairs. The tail is used as a counterbalance to keep the animal from tumbling over as it leaps through the air. Flying squirrels that glide on folds of skin between their arms and legs have a fluffy tail. As well as for balance, the tail is used as an air brake and rudder during glides. Hamsters have only a small stumpy tail, and cavies have no obvious tail at all.

▼ *To climb safely down this grass stem, the harvest mouse uses its prehensile tail. The prehensile tail is curled around the stem, helping the mouse to balance and providing some extra grip.*

Nervous system

A rat's nervous system provides the link between the outside world and the animal's body. Useful information about the environment is collected by the animal's senses. A rat has five senses—sight, hearing, touch, taste, and smell.

Changes detected in the environment are transmitted as electric pulses along sensory nerves to the central nervous system (CNS). Nerves are bundles of branched cells called neurons. The CNS is made up of the brain and the spinal cord—a mass of nerve cells running down the back inside the vertebrae. The CNS interprets all the sensory information and transmits more signals along another set of neurons, called motor neurons, to so-called effectors. Effectors are body structures that respond to changes in the environment. Obvious examples of effectors are muscles that move the body away from danger or toward food. However, motor neurons also transmit to various glands and body organs. These effectors may release hormones and produce longer-term changes, such as changes in fertility or growth.

▼ **Brown rat**
The brain and spinal cord constitute the central nervous system (CNS) of the rat. From the CNS, nerve fibers spread throughout the rat's body, forming the peripheral nervous system (PNS).

IN FOCUS

Communication

Brown rats live in large communities, or packs, of up to 200 individuals. A pack is dominated by the larger males, who communicate their dominance in a number of ways. These signals may be visual, such as threatening body and tail postures; or vocal, such as squeaks of alarm or recognition. Scratches and nips with the formidable teeth are also clear signals of which rat is in charge. Rats also communicate through scent. Rats of both sexes smear streaks of urine on all available surfaces. Males do this about 10 times as frequently as females, using fine hairs on the tip of their penis to leave their scented messages. The urine of an individual rat has a unique identifying odor. Male rats use their smell to broadcast their ownership of a certain territory. Females seek out the smell of their close female relatives.

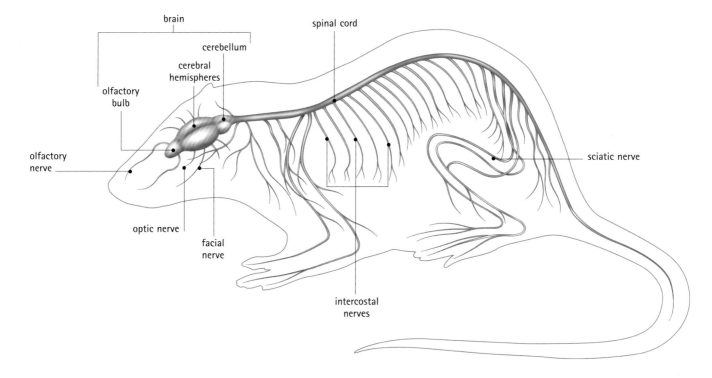

brain

cerebellum

spinal cord

cerebral
hemispheres

olfactory
bulb

olfactory
nerve

optic nerve

facial
nerve

intercostal
nerves

sciatic nerve

IN FOCUS

Lab rats

Compared with wild strains, brown rats used for experimentation are more docile, have larger eyes and ears, and have a longer tail. The white rat so familiar as a lab animal is a domesticated breed of brown rat. These rats are used for studying the living mammalian body under a range of conditions. They are used to test the effects of drugs, cosmetics, and many other substances that people might come into contact with. One feature that scientists exploit is the rat's ability to learn and remember things. One of the most famous tests involves putting a rat in a maze and observing how quickly it learns the layout. If this is done with a large number of rats, it provides a measure of their average intelligence. Researchers then test how certain drugs or even genetic changes can affect the rats' intelligence.

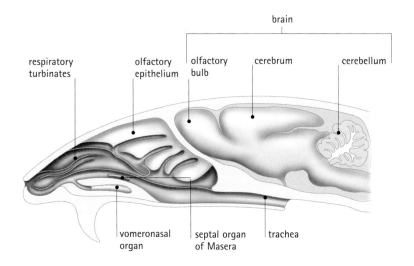

respiratory turbinates • olfactory epithelium • olfactory bulb • brain (cerebrum, cerebellum) • vomeronasal organ • septal organ of Masera • trachea

Senses

A brown rat's sense of sight is not very good, as is typical of rodents that spend most of their time in the dark. Its small eyes are on either side of its head and so the rat does not have binocular vision—the type of vision that occurs when the two fields of vision overlap to produce a sharp three-dimensional image. Instead, rat eyes are farsighted and are able to scan a wide area looking for movements.

Rats rely much more on their other senses. Their tubular outer ears capture sounds very effectively, although for a mouselike rodent, rat ears are quite small compared with its body size. The long tail is used to feel objects to the side and above the rat, as well as provide early warning of attack from behind. The long whiskers sticking out the side of the rat's snout are also sensitive touch organs, as are the soles of the forepaws. Both, for example, can pick up vibrations in the ground.

By far the most acute of a brown rat's sense organs is its nose. The pointed snout creates a long nasal cavity, which is lined with odor-sensitive cells. The brown rat's excellent sense of smell increases its ability to find food and avoid predators, and is one of the main reasons this animal has flourished and become so common.

▲ SENSE OF SMELL
The brown rat's sharpest sense is that of smell. Odor-sensitive cells line the long nasal cavity (olfactory epithelium). A large part of the rat's brain— the olfactory bulb— is dedicated to interpreting smell.

▼ *The rat's nervous system constantly monitors position and balance, enabling the animal to walk safely across this rope.*

Circulatory and respiratory systems

Brown rats are warm-blooded animals: that is, they maintain their body temperature within a narrow range of temperatures. The energy needed to do this is released from food. In cold conditions, energy is expended to warm the animal; in hot conditions, energy is expended to cool the animal. So-called cold-blooded animals, such as reptiles, do not have this ability. Instead, their body temperature rises and falls with the temperature of their surroundings. Because rats are warm-blooded, they are not limited by changes in temperature and can exploit habitats in all but the most severe of climates.

Oxygen

Warm-blooded animals use a lot of energy and require a lot of oxygen to release this energy from their food. Rats get their oxygen by breathing in air through their nose. The air is drawn down into the lungs by a temporary decrease in pressure inside them. This is produced by expanding their volume using rib muscles and the diaphragm—a large dome-shaped muscle attached to the lower rib cage. These muscles work together to create the

▼ **Brown rat**
The circulation and respiratory system provides the body's cells with the respiratory gases and broken-down food products that the cells need for the processes of respiration.

IN FOCUS

Size and climate

There is a link between the size of some wide-ranging mammal and bird species and where they live. Bergmann's rule, named for the German zoologist Karl Bergmann, states that for warm-blooded animals (such as mammals and birds) the largest members of a particular species will tend to live in colder regions. Therefore the largest brown rats are found living on the fringes of the Arctic tundra. This size difference is connected with the ratio of a mammal's surface area to its volume. As a body grows, its ratio of surface area to volume decreases. For this reason, larger warm-blooded animals loose heat less quickly than smaller ones and therefore need proportionally less food to survive. In cold climates, keeping warm is easier for larger mammals than for smaller mammals and consequently larger animals of a particular species tend to be more common.

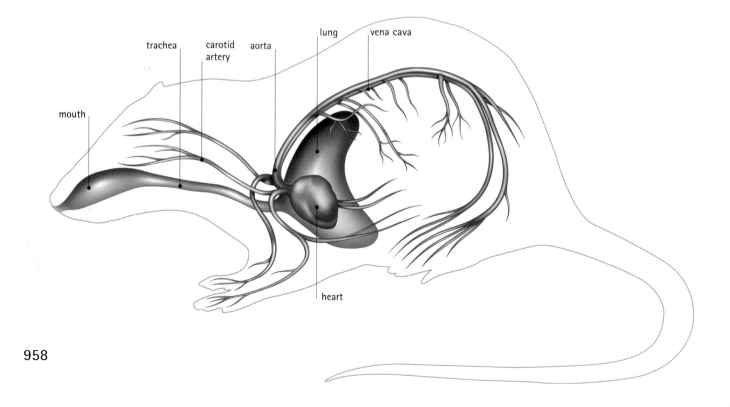

heaving motion of the chest associated with breathing. Since pressure is lower inside the enlarged lungs than outside, air rushes in and restores the balance. Once inside the lungs, oxygen passes into the blood vessels that surround the lungs. The oxygen gas combines with hemoglobin, a red iron-containing protein inside red blood cells. Carbon dioxide gas, a waste product of metabolism, passes from the blood into the air in the lungs. This gas is then breathed out.

Blood

A rat's blood is a liquid that contains a variety of cells and dissolved chemicals. Its function is to carry things body cells need, such as oxygen, nutrients, and hormones (chemical messengers) around the body. Blood is pumped through tubes, or vessels, by a single heart. The vessels that carry blood away from the heart are called arteries. Those that bring it back again are veins. A rat's blood supply is typical of other placental mammals.

GENETICS

Stem cells

Blood cells need to be frequently replaced. Red blood cells and most white blood cells are produced in the bone marrow, located at the center of large bones. All new body cells, including blood cells, are formed from stem cells. These are cells whose function is not yet decided and which retain the ability to differentiate into a variety of final types. Cell differentiation occurs when certain genes in the cell are switched and the cell begins to perform a specific role. Once it is

differentiated, the cell cannot be transformed into another form.

The stem cells in blood marrow are unipotent. This sort of stem cell can divide into only a limited number of cell types—in this case, blood cells. Pluripotent stem cells can divide into many types. The stem cells of a developing embryo may be able to differentiate into all types of cell.

One of the main fields of biotechnology is stem cell research. Doctors think that pluripotent stem

cells could be used to repair body parts that have become damaged. For example, stem cells could be grown into nerve cells to reconnect a paralyzed person's damaged spinal column.

Human stem cell research is highly controversial because it would involve cloning human embryos for the sole purpose of harvesting their stem cells. Using human embryos in this way is currently illegal in the United States. Instead lab rats and other animals are used for controversial research.

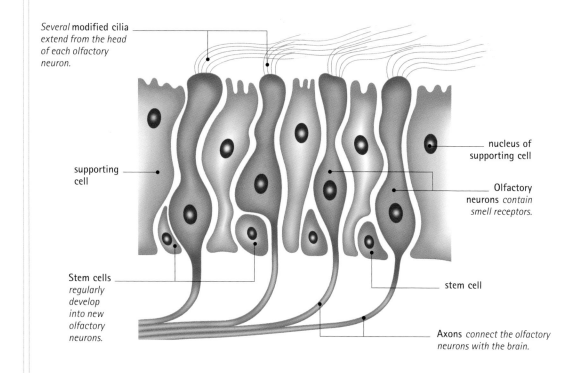

Several modified cilia extend from the head of each olfactory neuron.

supporting cell

Stem cells *regularly develop into new olfactory neurons.*

nucleus of supporting cell

Olfactory neurons *contain smell receptors.*

stem cell

Axons *connect the olfactory neurons with the brain.*

◀ STEM CELLS IN OLFACTORY EPITHELIUM
Mammal
Not all stem cells produce new blood cells. In the nose of a rat or other mammal, there are ciliated neurons specialized for sensing smell. These neurons are short-lived, lasting about one month before they are replaced by stem cells.

Digestive and excretory systems

Brown rats are omnivores. They will eat almost anything, including leather, wax, and soap. However, they prefer to feed on meat if they can get it. A brown rat needs to eat as much as one-third of its body weight every day, so it is constantly on the lookout for food. Rats are largely scavengers, but they also prey on small animals, such as bird chicks, mice, frogs, and even other rats.

Outside of natural habitats, brown rats depend on discarded human food. There are probably more brown rats in the world's sewers than in their original habitats—forests and wetlands. This association with people has encouraged the rodents to live in areas that would otherwise not have supported them.

Equipped for foraging

A brown rat's body makes it an expert forager. It is athletic, able to swim across water and climb around and jump over obstacles in order to investigate every nook and cranny for morsels of food. Highly developed senses of

▼ **Brown rat**
The digestive system of the brown rat has evolved in such a way that these animals are able eat a very wide range of food types.

Sick to death

Rats are physically incapable of vomiting. Therefore, if they eat something toxic, they are highly susceptible to being poisoned as they are unable to expel the harmful material. As a result, they are extremely cautious when first faced with an unfamiliar food. Usually one or two individuals in a pack will taste it, then wait awhile. If it makes them feel unwell, they, and other pack members know to avoid it in future. Biologists call this wariness neophobia, and it is one of the reasons that agriculturalists and pest control organizations are constantly seeking novel types of poison. The most effective poisons have a delayed effect. If a rat feels no immediate ill effects, it might be tempted to go back for a lethal dose.

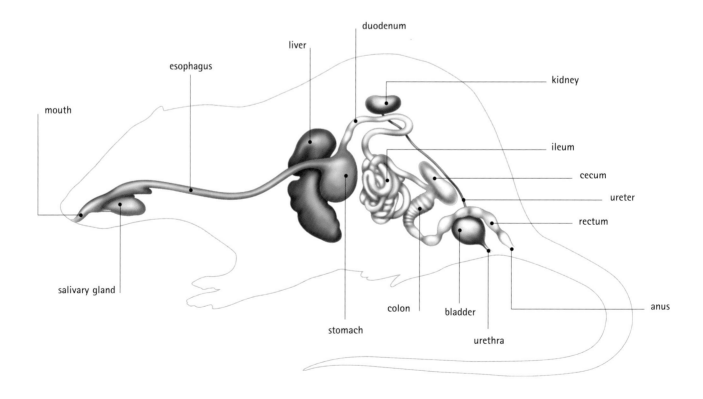

duodenum
liver
esophagus
kidney
mouth
ileum
cecum
ureter
rectum
salivary gland
colon
bladder
anus
stomach
urethra

COMPARATIVE ANATOMY

Nothing wasted

Although few mouselike rodents have as varied a diet as brown rats, many eat a wide range of foods. For example the wood mouse and Eurasian harvest mouse live in grassy habitats and eat a variety of seeds, fruits, and cereal grains. They supplement this diet with insect larvae, beetles, and grasshoppers when the opportunity presents itself, but most of their food comes from plants. Plant foods contain tough cellulose fibers, which are hard to digest. As is the case for many other rodents, the small intestines of the harvest and wood mice cannot break down the cellulose and absorb its nutrients. Instead, bacteria in the large intestine, or cecum, digest the cellulose into sugars that can be absorbed by the body. However, only the small intestine can absorb food into the blood. Therefore the mice eat their own feces so they can remove as many nutrients from a meal as possible. As a result of this process of coprophagy (eating feces), 80 percent of the mouse's food is absorbed.

▶ Brown rats have a very wide diet and predate small animals such as frogs and birds.

▼ This giant pouched rat is raiding a fruit store. The common name of these rats comes from their large size and the cheek pouches in which they store food.

smell and touch allow rats to forage in total darkness. Brown rats are able to build up an accurate mental map of their surroundings, as shown by their ability to escape from mazes in laboratories. This mental map lets them move around complex sewer systems or burrows and therefore search for food in an efficient manner.

Once a rat finds something edible, its teeth can cut through all but the toughest items. Rats store food in their nests. By doing this, brown rats are important seed dispersers. They carry fruits and nuts away from plants and stash them in another location. Often the rats do not eat all the food in their store, and the seeds sprout into a new plant.

Reproductive system

COMPARE the reproductive strategy of a rat with that of a **WOLF**. In wolf packs, only a dominant pair produce young. The other wolves help raise this pair's offspring.

COMPARE the undeveloped (altricial) nature of rat pups with the state of development of a newborn **ELEPHANT**. Elephant calves are precocious at birth; that is, they are able to walk, see, and have a degree of independence.

Brown rats live short, busy lives in which death is an ever-present threat. When living in captivity, rats can reach the age of five years. However, in the wild a brown rat will be lucky to reach its second birthday without being eaten by one of its many predators, or being struck down by one of the many and varied dangers present in an urban environment. Therefore, as with other mouselike rodents, the brown rat's strategy for success is to breed early, prolifically, and often. Unlike longer-lived mammals that nurture a small number of young, investing them with the resources needed for survival, rats produce large numbers of young with minimum effort. The brown rats' high rate of reproduction is another of the reasons that these rodents have been able to proliferate across the world.

Breeding machines

A female brown rat can produce between 2 and 22 pups in a single litter, and can have up to 12 litters in a year. Even an average rat will give birth to 60 young a year, and some produce twice that number. Female brown rats become sexually mature at the age of two or three months. Male rats mature a few weeks earlier, but they are unlikely to be able to compete for mates with older and larger males until they have grown a bit more. If all offspring survived and bred to their full potential, a single pair of rats could give rise to 15,000 descendants within a year.

Both male and female brown rats have many mates. A female rat is receptive to mating for a 20-hour estrous period. During this time, she may mate 500 times with different males. The males compete for access to females, forming a dominance hierarchy based on body size. Although mating takes place all year, most pups are born in the warmer months.

▼ UROGENITAL ORGANS
Brown rat

The main features of the reproductive system of the male and female brown rat are typical of those of most mammals. The diagrams also show some elements of the urinary and excretory systems.

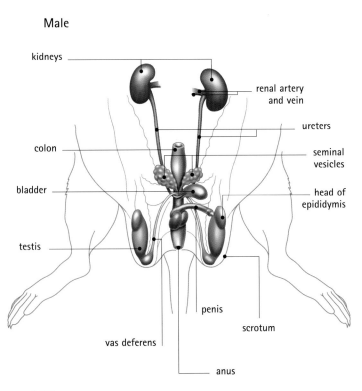

Male

- kidneys
- colon
- bladder
- testis
- renal artery and vein
- ureters
- seminal vesicles
- head of epididymis
- penis
- scrotum
- vas deferens
- anus

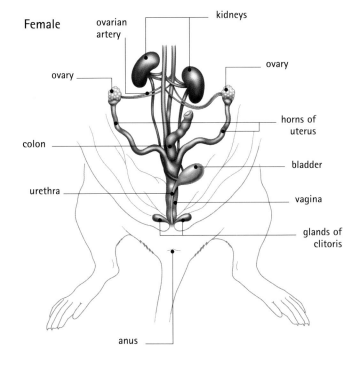

Female

- ovarian artery
- kidneys
- ovary
- ovary
- colon
- horns of uterus
- urethra
- bladder
- vagina
- glands of clitoris
- anus

Passing messages

The smell of urine has an important effect on the reproductive systems of female rats. When a juvenile female rat smells the urine of an unrelated male rat, she begins to sexually mature more quickly. This allows her to take advantage of breeding with an unrelated male, which results in stronger offspring than mating with a closer relative. Also, if a young female smells the urine of a pregnant rat or one that is suckling young, she will mature more quickly. With the older rat unavailable for mating, the young rat will find more mates. However, if older female rats are not producing young, the smell of their urine delays younger females from reaching puberty. This delay reduces the competition for mates in areas overcrowded with rats.

Sexually mature female rats respond to these smells in a similar way. They prepare for pregnancy when they smell new males or pregnant females. However, they may stop ovulating altogether if there are several other adult females close by. In the incestuous world of a brown rat pack, the opportunity to mate

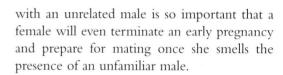

CLOSE-UP

Life cycle of the harvest mouse

The high rate of reproduction of brown rats is mirrored in life cycles across the mouse family. For example, baby harvest mice are born after a gestation period of just 18 days. There are usually about four or five young in each litter. Each one is less than 0.75 inch (20 mm) long and weighs just 0.025 ounce (0.7 g). They are helpless at birth. At the age of 12 days, however, the juveniles' eyes have opened and they are able to move about in the nest. Sixteen days after birth, the mother stops suckling the pups and they become independent. By this time their mother is already pregnant with the next litter.

with an unrelated male is so important that a female will even terminate an early pregnancy and prepare for mating once she smells the presence of an unfamiliar male.

Communal nests

Brown rats are pregnant for about 23 days. Newborns weigh 0.17 ounce (5 g). They are hairless and their eyes remain closed until they are about 17 days old. Young that are born in such an underdeveloped and helpless state are described as altricial. Female brown rats have six pairs of mammary glands, and the pups are fed exclusively on milk for about four weeks. They develop quickly, and a few weeks after weaning the pups leave their mother's care for good. Just 18 hours after giving birth, a female rat's uterus begins preparing for the next pregnancy, and she is ready to give birth again soon after weaning

Female rats seek out the company of closely related females, and all their pups are often raised in large communal nests. Brown rats are cooperative breeders. A female will suckle another rat's young and will continue to care for them if the true mother dies. Male brown rats play no part in caring for the young.

TOM JACKSON

▼ *Newborn brown rats are very undeveloped and depend completely on their mother.*

FURTHER READING AND RESEARCH
Alderton, David. 1996. *Rodents of the World.* Facts On File: New York.
Conniff, Richard. 2002. *Rats! The Good, the Bad, and the Ugly.* Crown Publishers: New York.
Paxinos, George (ed.). 2004. *The Rat Nervous System.* Elsevier Academic Press: Boston, MA.

Red deer

ORDER: **Artiodactyla** FAMILY: **Cervidae**
SPECIES AND SPECIES: *Cervus elaphus*

There are about 41 species of true deer in the family Cervidae. Deer live in a wide range of habitats, from tropical forests to Arctic tundra, and they range in size from the diminutive pudu, which reaches a height of 14 inches (35 cm), to the moose, which grows to 7.5 feet (2.3 m) tall. The red deer is the largest species of the genus *Cervus* and is widespread throughout the northern hemisphere.

Anatomy and taxonomy

Scientists categorize all organisms into taxonomic groups based on anatomical, biochemical, and genetic similarities and differences. The 41 species of deer are characterized mainly by the presence of antlers in the males.

- **Animals** are multicellular and gain their food supplies by consuming other organisms. Animals differ from other multicellular life-forms in their ability to move from one place to another (in most cases, using muscles). They generally react rapidly to touch, light, and other stimuli.

- **Chordates** At some time in its life cycle a chordate has a stiff, dorsal (back) supporting rod called the notochord that runs all or most of the length of the body.

▼ *Red deer belong to the true deer family, Cervidae, which contains about 41 species in 17 genera. Males of all Cervidae species possess antlers: hornlike growths made of bone that are shed and regrown each year. Musk deer are placed in a separate family, Moschidae. Musk deer differ from true deer in lacking antlers and possessing an abdominal musk gland.*

Animals
KINGDOM Animalia

Chordates
PHYLUM Chordata

Vertebrates
SUBPHYLUM Vertebrata

Mammals
CLASS Mammalia

Even-toed ungulates
ORDER Artiodactyla

Pigs and hippos
SUBORDER Suiformes

Ruminants
SUBORDER Ruminantia

Camels and llamas
SUBORDER Tylopoda

Cattle, antelope, goats, and relatives
SUPERFAMILY Bovoidea

Deer, pronghorns, and musk deer
SUPERFAMILY Cervoidea

Giraffes and relatives
SUPERFAMILY Giraffoidea

True deer
FAMILY Cervidae

Red deer
GENUS AND SPECIES
Cervus elaphus

Fallow deer
GENUS AND SPECIES
Dama dama

Roe deer
GENUS AND SPECIES
Capreolus capreolus

White-tailed and mule deer
GENUS *Odocoieus*

Muntjac deer
GENUS *Muntiacus*

Caribou, or reindeer
GENUS AND SPECIES
Rangifer tarandus

● **Vertebrates** In vertebrates the notochord develops into a backbone (spine or vertebral column) made up of units called vertebrae. The vertebrate muscular system that moves the head, trunk, and limbs consists primarily of muscles that are bilaterally symmetrical—they are arranged in mirror-image fashion on either side of the backbone.

● **Mammals** Mammals are warm-blooded vertebrates that have hair made of keratin. Females have mammary glands that produce milk to feed their young. In mammals, the

▲ *Male red deer have large antlers made of bone, which are shed each year after mating and then regrown the following year.*

lower jaw is a single bone, the dentary, hinged directly to the skull, a different arrangement from that found in other vertebrates. A mammal's inner ear contains three small bones (ear ossicles), two of which originated as the jaw mechanism of mammalian ancestors. Mammalian red blood cells, when mature, lack a nucleus; all other vertebrates have red blood cells that contain nuclei.

FEATURED SYSTEMS

EXTERNAL ANATOMY Red deer are long-legged, with a short tail and a coat of coarse brown hair. Every year, males grow a new set of branching antlers made of bone. *See pages 967–969.*

SKELETAL SYSTEM Deer are built to run, with long leg and foot bones and just two functional toes on each foot. *See pages 970–972.*

MUSCULAR SYSTEM There are large blocks of striated muscle associated with the shoulders, neck, and rump. *See pages 973–974.*

NERVOUS SYSTEM Deer are moderately intelligent, with a range of acute senses. Smell, hearing, and vision are excellent. *See pages 975–976.*

CIRCULATORY AND RESPIRATORY SYSTEMS The lungs are large, as befits a mammal adapted for running. Double circulation is maintained by a four-chamber heart. *See pages 977–978.*

DIGESTIVE AND EXCRETORY SYSTEMS Deer are generalist herbivores with a ruminant digestive system able to cope with all kinds of plant material, from tender new leaves to twigs and bark stripped from trees. *See pages 979–980.*

REPRODUCTIVE SYSTEM Deer have an annual sexual cycle that influences all other body systems. Males use their antlers to fight each other for reproduction rights. *See pages 981–983.*

◀ *The antlers of a male fallow deer vary in appearance according to age. Up to three or four years, males produce antlers with spikes, but from three years onward the male's antlers may develop broad "palmate" areas as shown here.*

● **Artiodactyls** Even-toed ungulates have two or four toes on each foot. In all but the hippopotamuses, only the third and fourth digits are weight-bearing. In addition to true deer, the group includes cattle, sheep, antelope, camels, llamas, pigs, peccaries, giraffes, hippos, the pronghorn, mouse deer, and musk deer.

● **True deer** True deer are slender, long-legged animals, with cloven hooves; and males of all but one species (the Chinese water deer) bear antlers. Caribou (called reindeer in Europe) are unusual in that both males and females have antlers. Antlers are temporary structures made of bone, which are shed and regrown every year. Male deer are usually larger than females, and they tend to put on weight in preparation for the breeding season, or rut, when competition between males becomes intense, with much posturing and aggressive behavior.

The smallest of all deer, pudus, are native to South America. These are secretive forest dwellers, with a short body and short, slender legs. Males grow short spikes for antlers. Roe deer are widespread throughout Europe and Asia. These small, graceful forest specialists live alone or in small groups. The males' antlers have distinctive rough bases but rarely acquire more than three points. The moose (known in Europe as the elk) is the largest living deer.

Most deer can swim if need be, but moose are especially at home in water and spend much of their time wading in pools and swamps in search of aquatic vegetation. The enormous palmate antlers of a large male can span more than 6 feet (2 m). Caribou are the most northerly dwelling and cold-adapted of all deer. They have a thick shaggy coat and large spreading hooves that spread their weight and enable them to walk on snow. They live in large herds, and both males and females grow antlers. The white-tailed deer is common and widespread from southern Canada to northern South America. It is an elegantly proportioned deer, with a reddish brown summer coat, fading to dull gray-brown in winter. Males grow large branching antlers.

● **Placental mammals** nourish their unborn young through a placenta, a temporary organ that forms in the mother's uterus during pregnancy.

● **Ungulates** Hoofed mammals, or ungulates, include most of the world's large grazing and browsing herbivores, such as odd-toed ungulates (horses, rhinos, and tapirs of the order Perissodactyla) as well as cattle and deer. All hoofed mammals have limbs adapted for running, with nails modified into hard hooves. Ungulates have elongated foot bones and a reduced number of digits (fingers and toes), and the thumb and big toe are absent.

● **Red deer** This species of true deer is widespread in the northern hemisphere, although its size may vary according to local circumstances. In winter the coat is dark brown, but in summer it becomes tan. Males are distinguished by their large antlers, which are replaced yearly and which they use in battles over reproductive rights.

External anatomy

CONNECTIONS

COMPARE the antlers of the red deer with the horns of the **WILDEBEEST** and the **RHINOCEROS**. Horns are made mostly of keratin, with a core of bone, and grow continuously throughout the animal's life. Antlers are solid bone and are shed and regrown every year.

Deer are an anatomically and ecologically conservative group. In other words, they are all built along more or less similar lines and have broadly similar lifestyles.

Despite this general similarity deer vary widely in size. In the case of red deer, they also vary considerably in stature from place to place. This variability reflects the species' broad distribution and the wide range of habitats to which local races have adapted. Some people think the different races should be split into several species to reflect their variability.

Deer, antelope, and gazelles

Deer are graceful animals, but more robust than their cousins the gazelles and antelope. Most deer are built for the cool climates of higher latitudes—they need extra bulk to help them keep warm in winter and survive seasonal food shortages. Antelope, however, are restricted to tropical regions, where keeping warm is less of a problem and evading predators is a major priority. As a result they tend to be less robust but exceptionally agile.

▼ Moose are the largest of all the deer and, like fallow deer, they have palmate antlers. A flap of skin called a bell, or dewlap, grows under the moose's chin. The bell grows with age and probably helps the animal disperse the scent of urine.

GENETICS

Red deer, elk, or wapiti?

The red deer is one of the world's most widespread mammals. It is found throughout the northern hemisphere, and it is known by different names in different locations. In North America the red deer was widely known as elk, but this caused confusion with the European elk, or moose. Now, the old Shawnee Indian name *wapiti* is preferred for red deer. There are marked differences between red deer populations in different parts of the species' range. Wapiti are generally large, whereas Scottish red deer, which live on open moorland, are relatively small. These differences are sufficient to warrant the classification of red deer in several subspecies, which may in time become full species.

Nevertheless, deer and antelope share many features, not least long, slender legs and great athleticism. Both of these characteristics are adaptations for escaping from predators. Like antelope, deer are alert, wary animals, with large, mobile ears that swivel this way and that to pick up small sounds that might signal danger. Their eyes are large and positioned on the side of the head, providing a wide field of vision. The nostrils are at the tip of the snout and are surrounded by a large area of moist black skin called the rhinarium. Many deer

▼ *During the winter, the red deer's reddish coat turns grayish brown, with lighter patches on the rump and underside.*

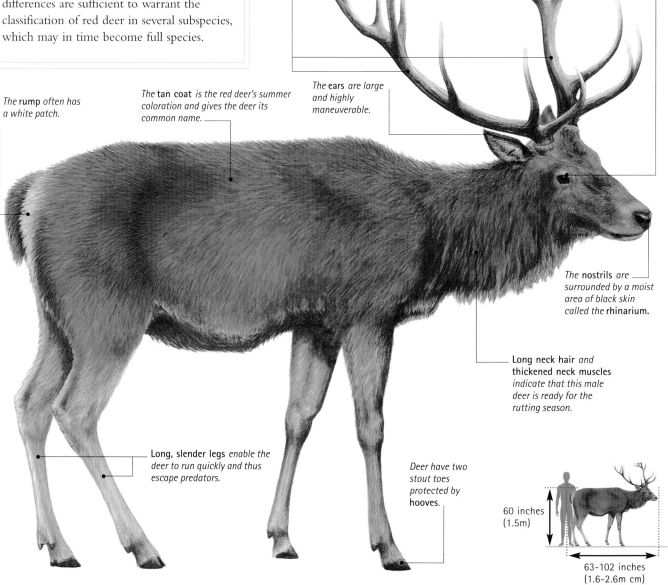

The **antlers** *are made of bone and have pointed branches called tines.*

tines

Large eyes *placed at the side of the head help the deer to detect predators.*

The **rump** *often has a white patch.*

The **tan coat** *is the red deer's summer coloration and gives the deer its common name.*

The **ears** *are large and highly maneuverable.*

The **nostrils** *are surrounded by a moist area of black skin called the* **rhinarium.**

Long neck hair *and* **thickened neck muscles** *indicate that this male deer is ready for the rutting season.*

Long, slender legs *enable the deer to run quickly and thus escape predators.*

Deer have two stout toes protected by hooves.

60 inches (1.5m)

63–102 inches (1.6–2.6m cm)

COMPARATIVE ANATOMY

Choose your weapon

The antlers of true deer come in all shapes and sizes, but one species, the Chinese water deer, lacks antlers completely. Instead, both males and females have upper canine teeth modified into a pair of sharp tusks—these are best developed in males. Other tusked deer include the muntjacs, the musk deer of the family Moschidae, and the lesser Malay mouse deer.

▶ *Male lesser Malay mouse deer grow a pair of small tusklike upper canine teeth, which protrude from the mouth.*

have facial glands that produce small quantities of a secretion containing powerful scents and pheromones. This is wiped on other deer or inanimate objects to advertise information about the individual deer that produced it.

Forest life

Most deer are forest dwellers, and this fact is reflected in their body shape and coloring. The body is narrow and tapers slightly toward the front, making it easier for these animals to pass through gaps in closely spaced vegetation. Many species (although not the red deer) are patterned with pale spots that mimic dappled sunlight and break up the body's outline. Spots are most common in young deer, for whom effective camouflage is especially valuable. Adult deer molt twice a year, in spring and fall. The winter coat of the red deer is thick and shaggy and a rather drab shade of grayish brown. In contrast, the summer coat is short, sleek, and a warm russet color—the color that gives them their English common name.

◀ ANTLERS
Different deer have widely different types of antlers. In red brocket deer the antlers are generally just simple spikes; pampas deer have short antlers with several tines; and moose have large palmate antlers with many tines.

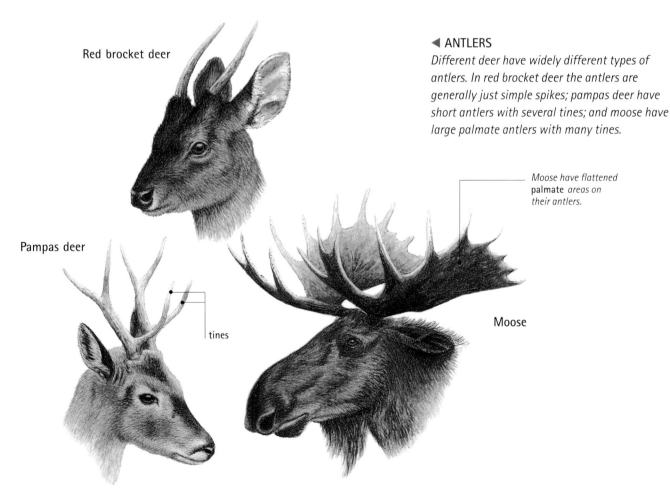

Moose have flattened **palmate** *areas on their antlers.*

Red brocket deer

Pampas deer

tines

Moose

969

Skeletal system

Deer have long, powerful legs, owing more to long foot bones than to long leg bones. Deer stand and walk on the very tips of their toes, using a stance that zoologists call digitigrade. As with other grazing ungulates, long legs necessitate having a long neck: the long neck enables the animal to feed without having to crouch.

The limb bones are adapted for strength and speed. In each leg, the smaller of the lower limb bones (the ulna and fibula) are short, limiting the degree to which the limb can rotate—deer cannot turn their feet out to the sides. The third and fourth foot bones, the metacarpals and metatarsals, are long and fused to form a cannon bone, while those on either side (the second and fifth bones of each foot) are present but serve little function. The first digit (equivalent to the thumb or big toe in other mammals) is absent, and the second and fifth digits are small. As in cattle and antelope, only the third and fourth digits are developed as functional weight-bearing toes.

The skull is robust, and tapers to a narrow snout with a large cavity to accommodate the large nasal cavity, which contains a sensitive smelling apparatus. The orbits (eye sockets) and the lower jaw are also large.

▼ Red deer

The prominent feature of the red deer's skeleton is the long limbs, which are suited to fast running. In addition, note the two pecidels, from which the antlers grow, and the short ulnas and fibulas.

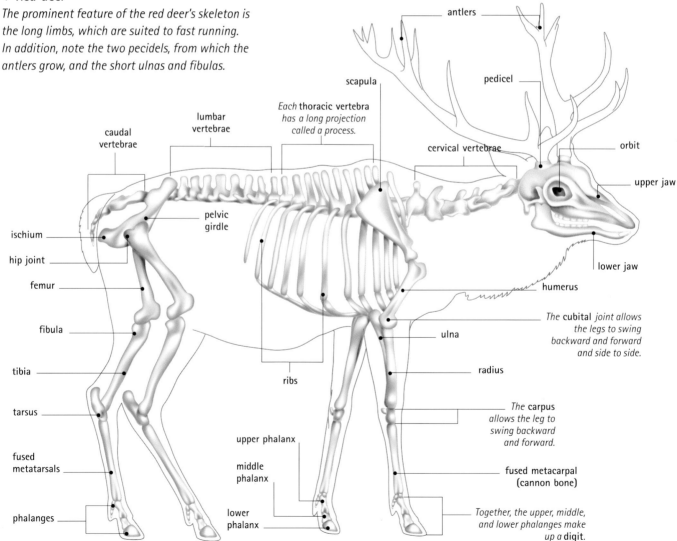

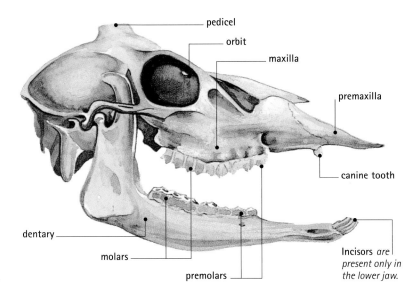

pedicel

orbit

maxilla

premaxilla

canine tooth

dentary

molars

premolars

Incisors *are present only in the lower jaw.*

Natural healing

At cellular level, the regrowth of antlers has much in common with the healing process. What biologists have not yet been able to work out is how the regrowing tissues are able to form such precise copies of the bone that existed before. Biologists think that if they can understand this process, there might be some way the knowledge can be used medically—for example to help amputees regrow lost body parts.

Teeth

Red deer have three pairs of incisors in the lower jaw and six pairs of grinding cheek teeth (three pairs of premolars and three pairs of molars) in each jaw. The incisors are used for cropping vegetation; they function by biting against the upper gum. There are no front teeth in the upper jaw. Most red deer also have a pair of small canine teeth in the upper jaw. In certain other species of deer the canines grow into tusks, but in red deer they are redundant—and in some individuals they never appear at all.

▲ SKULL

The red deer's dentition is typical of that of a ruminant: there are only two small canines, and the molars and premolars are covered with ridges that enable the deer to grind tough plant matter.

Headgear

Male deer have bony outgrowths of the skull just above the eyes. These are called pedicels, and it is from these that antlers grow each year. Unlike antlers, pedicels are permanent features of the skull.

When a stag casts his antlers after the rut, the wound bleeds freely for a few minutes before the blood clots to form a scab. Like most bones, those of a deer's skull are wrapped in fibrous connective tissue called periosteum (meaning "around the bone"). The periosteum of the

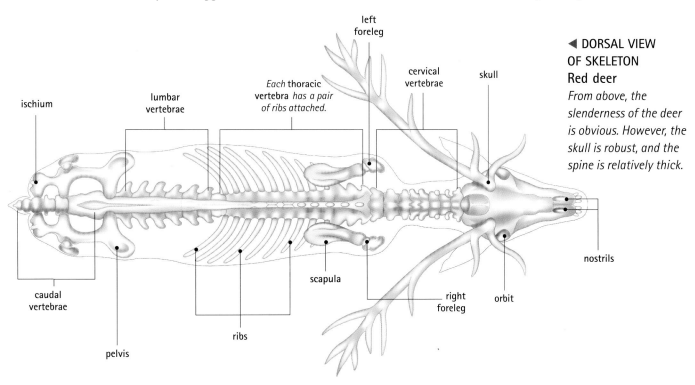

left foreleg

cervical vertebrae

skull

Each thoracic vertebra *has a pair of ribs attached.*

lumbar vertebrae

ischium

caudal vertebrae

pelvis

ribs

scapula

right foreleg

orbit

nostrils

◀ DORSAL VIEW OF SKELETON
Red deer
From above, the slenderness of the deer is obvious. However, the skull is robust, and the spine is relatively thick.

Antler

Horn

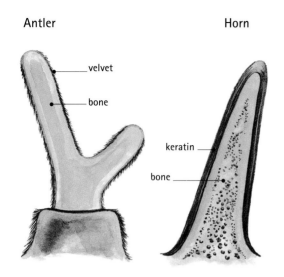

velvet

bone

keratin

bone

▶ *The blood on this caribou's antlers shows that the velvet has been shed recently. Caribou live in northern Canada, Alaska, Russia, and Scandinavia (where they are called reindeer). Both male and female caribou grow antlers.*

▲ HORN AND ANTLER

Animals such as rhinoceroses and bison have horns made from bone covered with a layer of keratin, which is the same material that forms fingernails, hair, and claws. Antlers are initially covered in a layer of furred skin called velvet, which falls off to reveal bare bone. Unlike horns, antlers fall off and are regrown each year.

pedicles (antler bases) contains a great many stem cells, which divide very rapidly to produce first cartilage, then bone. The cartilage forms a kind of scaffold, which in a matter of days and weeks is ossified (turned to bone). The antlers of large deer are the fastest-growing of all mammalian tissues. The antler of a well-fed red deer stag can grow up to 1 inch (2.5 cm) in a single day. Growing antlers

are made of living tissue—they are supplied with blood and nerves and are covered by a thin layer of skin with very short fur known as velvet. Developing antlers are sensitive to touch in the same way as other parts of the body, and they are warm. However, at the end of the growing season, the blood supply is cut off, the nerves die, and the velvet dries, splits, and falls away. All that is left is cold, hard bone.

CLOSE-UP

Growth of antlers

Biologists used to think that antlers grew at a steady rate, increasing in size each year according to a predictable pattern. The number of points on a pair of antlers was used to estimate the age of the stag. However, we now know the situation is not as simple as that. Young stags tend to have smaller antlers than old stags, but most reach a maximum at around six or seven years of age, after which there is little or no increase. Elderly stags, whose body is beginning to slow down, tend to have smaller antlers than they had during their prime.

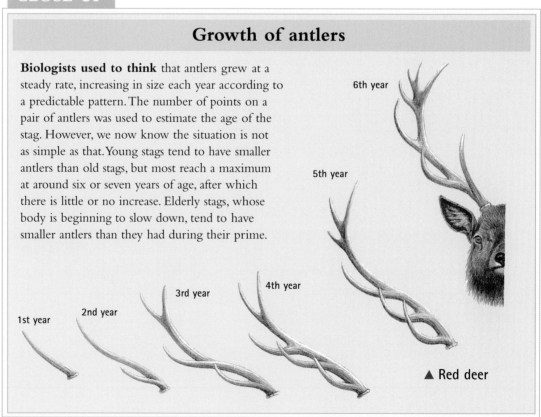

1st year

2nd year

3rd year

4th year

5th year

6th year

▲ Red deer

Muscular system

The musculature of red and other deer is broadly similar to that of most hoofed mammals. The largest muscle blocks are those associated with running and (in males) with fighting and supporting heavy antlers.

The front legs support more of the body's weight than the hind limbs, especially in males, whose antlers and massive neck and shoulder muscles make them very front-heavy. However, most of the propulsive power used in running comes from the hind limbs, which are therefore more muscular.

Deer muscles contain relatively large quantities of the pigment myoglobin, so deer meat, or venison, is dark in color. Red deer are large and powerful by deer standards but are still vulnerable to attack by large predators such as wolves. Red deer are adapted for speed, endurance, and agility as well as strength. As any athlete knows, these qualities are difficult to balance, since improving one usually means compromising the other. In nature there are sprinters and long-distance specialists, but relatively few animals excel in both areas. Red deer manage to combine the two surprisingly well.

Fast twitch and slow twitch

Physiologists now understand that the striated, or skeletal, muscles responsible for moving the body around come in two distinct types, known as fast-twitch and slow-twitch muscles.

COMPARE the leg musculature of the red deer with that of the *HUMAN*. Both are adapted for endurance running, but the deer can run much faster.

CONNECTIONS

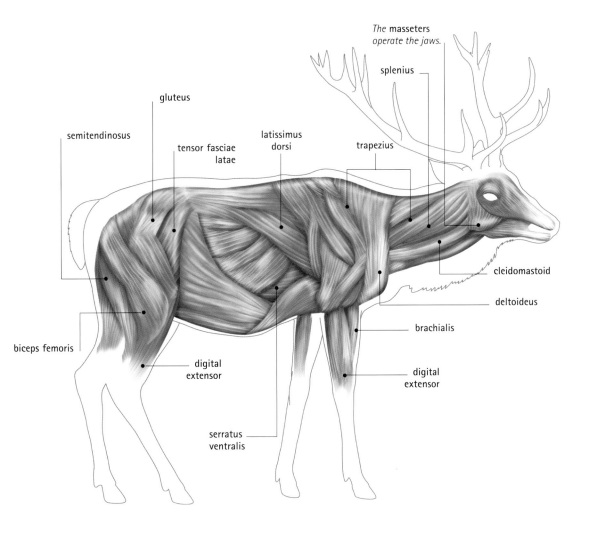

The masseters operate the jaws.

splenius

gluteus

semitendinosus

tensor fasciae latae

latissimus dorsi

trapezius

biceps femoris

digital extensor

serratus ventralis

cleidomastoid

deltoideus

brachialis

digital extensor

◀ **Red deer**
The red deer has particularly well-developed leg and neck muscles. The strong leg muscles enable the red deer to run fast, and the neck muscles support the weight of the antlers.

Body builders

In preparation for the annual breeding season known as the rut, mature stags undergo a remarkable physical change. The muscles of the neck, which will be used for wrestling rival stags, increase dramatically—the overall girth of the neck can double in a matter of weeks.

This amazing change is a result of increased production of the hormone (chemical messenger) testosterone. No wonder many human bodybuilders are tempted to use steroids—the same substance has a similar effect on human muscle mass.

▼ *Like other male deer, male fallow deer use their antlers to fight one another for reproduction rights. They lock antlers and use their strong neck muscles to try to twist their opponent's neck and shove him backward.*

Clicking reindeer

Reindeer make a clicking sound as they walk. The clicking is caused by a tendon rubbing across a bone in the foot and is a consequence of the arrangement of bones and muscles. The clicking enables an individual to sense the mood of the herd around it by listening to the clicks. If one set of clicks suddenly gets faster, it signals that a member of the herd has started moving quickly, possibly signaling danger. Instantly, every other deer in the herd is on the alert.

Most mammals have both types of muscles, in proportions that vary between species and between individuals. Sprinters have a greater proportion of fast-twitch muscle; endurance runners have more slow-twitch muscles.

The practice of "endurance hunting," in which trained human athletes literally run an animal such as a deer or antelope to death, was once widespread. Humans have the benefit of advanced communication skills, so such a hunt can be carefully planned; and because we run on two legs we can carry water and food to sustain us in the course of a long run. As long as the human runner can continue to harass the deer and prevent it from resting, feeding, or drinking, eventually it will collapse from exhaustion.

Nervous system

Like all vertebrates, deer have a central nervous system consisting of a brain and spinal cord. Nerves branch out from the spinal cord to all other parts of the body, forming a complex peripheral system that directs information gathered by the senses to the brain and sends signals to muscles and other organs. In male deer, at least part of the peripheral nervous system is lost every year—the furry skin that covers growing antlers contains nerves just like that covering the rest of the body. Growing antlers are sensitive to touch. The antler nerves wither and die each year as the velvet is shed but regrow the following year.

The structure of the brain tells the story of its own evolution. At the base of the brain are the areas that control the fundamental

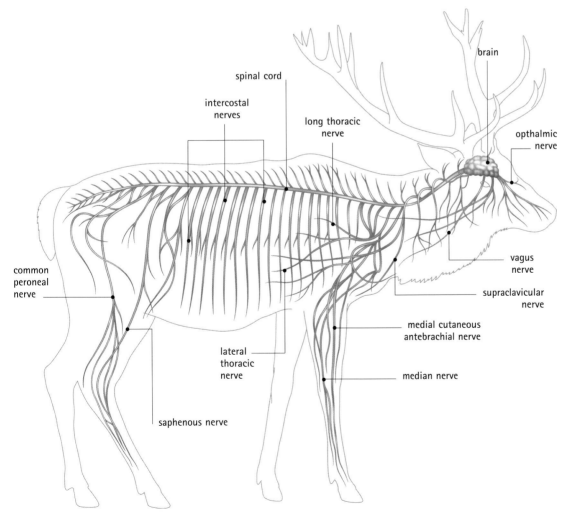

spinal cord

intercostal nerves

long thoracic nerve

brain

opthalmic nerve

vagus nerve

supraclavicular nerve

medial cutaneous antebrachial nerve

median nerve

lateral thoracic nerve

saphenous nerve

common peroneal nerve

◀ **Red deer**
A red deer's brain is relatively large, as are the brains of all cold-adapted northern species. The brains of tropical deer are relatively small. The difference is probably related to the greater demands of living in a seasonal climate, since each season places different demands on individuals to survive.

▼ *This male red deer curls back its lip and uses the vomeronasal organ in its mouth to sample chemicals in the air. Males use this organ to detect the sexual receptivity of females.*

CLOSE-UP

The vomeronasal organ and Flehmen

Like many other mammals, deer have an additional sense related to smell and taste but not quite the same as either. This sixth sense is associated with an organ in the roof of the mouth called the vomeronasal organ, or Jacobsen's organ. Sensitive receptors there can detect very small quantities of certain chemicals in the air, in particular those relating to the opposite sex. Male deer use a special behavior known as the flehmen response to sample the chemicals given off by female deer. Flehman involves curling the upper lip in an odd-looking grimace to draw air over the sensory cells.

processes essential to life. The forebrain is responsible for higher functions. At the very front is the cerebrum, consisting of two large, cerebral hemispheres. These are responsible for coordinating much of the animal's activity—including instinctive and learned behavior.

A large proportion of brain tissue is fat. The role of fat in the nervous system is principally to insulate the nerve axons, thus increasing the efficiency with which electrical impulses are transmitted.

Smell, vision, and hearing

Red deer have superb senses of smell, vision, and hearing and therefore are highly sensitive to their environment. Scent is important in many ways. The members of a population use scent to communicate, especially during the breeding season, when rival males spray urine onto their own belly and roll in urine-scented mud to create a powerful personal smell. Many deer have a pair of scent glands on the face just in front of the eye. These suborbital glands open up to release scent into the environment. Often this happens in conjunction with urination and during courtship. In red deer there are further scent glands on the lower hind legs and under the tail.

The senses of vision and hearing each play a major role in helping the deer avoid predation. The ears are mounted high on the head and can be swiveled almost continuously to detect sound in all directions. The large eyes are positioned at the side of the head and bulge slightly, thus providing the deer with a very wide angle of vision. Red deer have excellent night vision, and they also have color vision, although probably not as good as ours.

Circulatory and respiratory systems

Like all mammals, deer are warm-blooded: they maintain a constant warm body temperature. This creates optimum conditions for all the chemical processes that sustain life. Most deer are animals of temperate climates where the ambient temperature is usually well below their body temperature. Thus they have many adaptations that enable them to conserve body heat. Cool air enters the deer's nostrils and passes along convoluted nasal passages richly supplied with blood vessels.

Blood passing beneath the skin warms the air so it does not chill the lungs. The lining of the respiratory tract and in particular the delicate tissues of the lungs are moist, and by the time air is expelled from the lungs it has picked up a good deal of moisture. Some of this condenses on the insides of the nostrils on its way out, but if the deer is breathing heavily a good deal will be lost as vapor—the head of a bellowing stag, for example, is sometimes wreathed in steam.

▼ ARTERIAL AND CIRCULATORY SYSTEMS

As in other vertebrates, the red deer's heart pumps oxygen-rich blood around the body in arteries and oxygen-depleted blood to the lungs to pick up freshly inhaled air in veins (not shown). The two large lungs together can accommodate up to 5 gallons (20 l) of air.

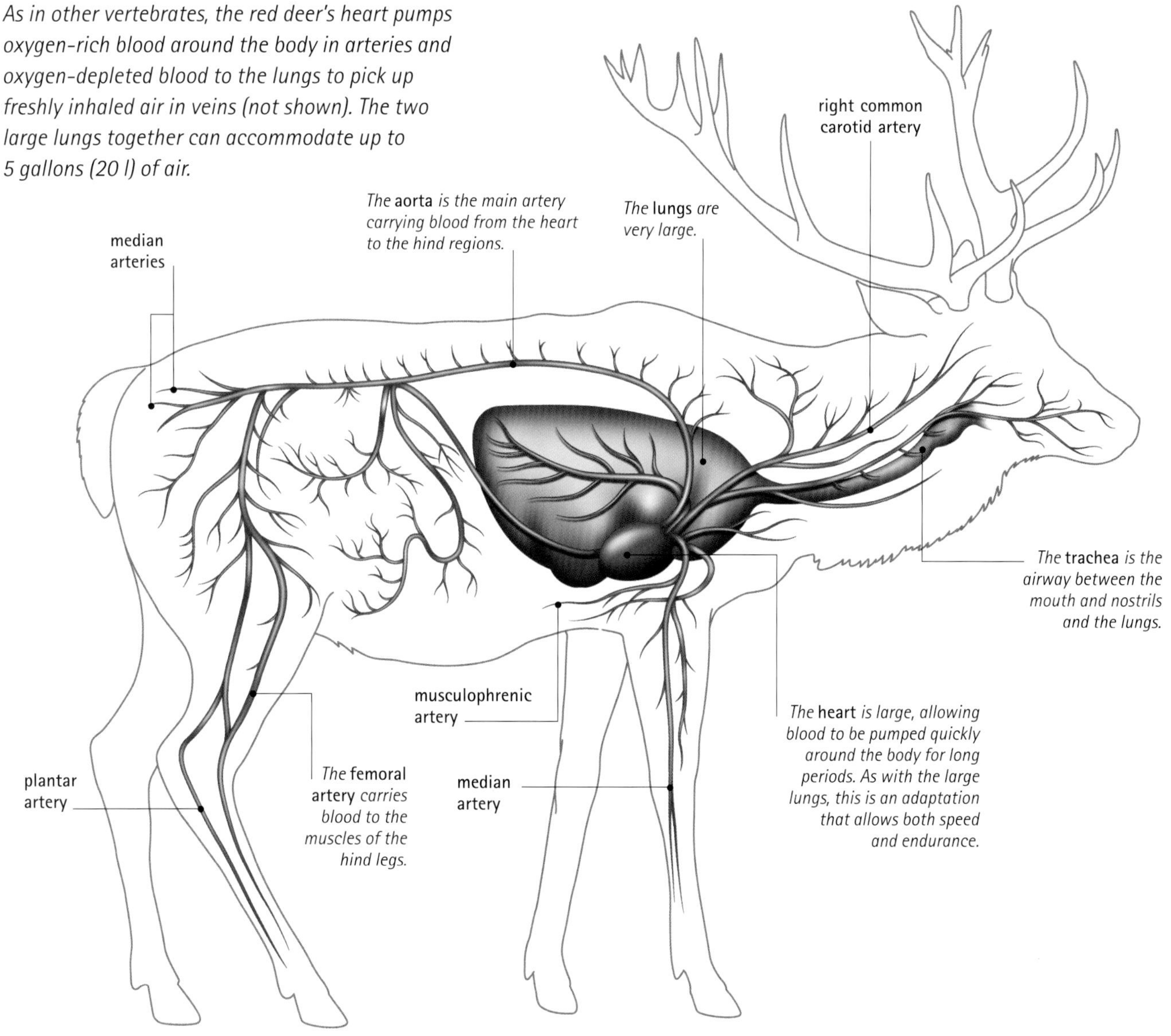

median arteries

The **aorta** *is the main artery carrying blood from the heart to the hind regions.*

The **lungs** *are very large.*

right common carotid artery

The **trachea** *is the airway between the mouth and nostrils and the lungs.*

musculophrenic artery

plantar artery

The **femoral** artery *carries blood to the muscles of the hind legs.*

median artery

The **heart** *is large, allowing blood to be pumped quickly around the body for long periods. As with the large lungs, this is an adaptation that allows both speed and endurance.*

Bellowing

Most true deer are silent during the mating season, but male red deer are very noisy. Stags bellow to advertise their control of a "harem" of females, or to challenge another stag's control. As two stags try to out-advertise each other, a female will often try to make a break for it. Then, the controlling stag will attempt to drive the unfaithful female back into his harem.

▶ *This male red deer is surrounded by a cloud of steam produced by bellowing. The male bellows to indicate his status and challenge other males.*

Lungs

Under normal circumstances red deer have a respiratory rate of between 20 and 35 breaths per minute. Red deer have a deep chest that accommodates large lungs. In a large individual the lungs have a volume of up to 5 gallons (20 l). Air enters through the trachea, or windpipe, which divides into two bronchi and then into a great many smaller branching tubes called bronchioles, terminating ultimately in a great many tiny chambers called alveoli. The walls of the alveoli are well supplied with blood vessels and provide a huge surface area over which gas exchange can take place.

Heart and blood

Oxygen taken up in the lungs is carried around the body bound to a pigment in the blood called hemoglobin. Hemoglobin releases oxygen when it reaches tissues where oxygen is needed for respiration.

The heart rate of a red deer varies greatly depending on activity, and also with age, sex, and levels of nutrition. The heart of a young animal beats considerably faster than that of an adult—typically 150 to 180 beats per minute in a resting newborn fawn, but as little as 40 beats per minute in a resting adult. As in all mammals, blood passes through a deer's heart twice on every circuit of the body, through the right-hand side on the way to the lungs and then through the larger left-hand side, which pumps it around the rest of the body.

Arteries and veins

Blood circulates to the extremities through arteries and veins that lie alongside each other. Arteries carry warm blood under high pressure outward from the heart. The blood traveling in the other direction in veins is under lower pressure and has usually been cooled as it passes close to the body surface. Because of the close proximity of veins and arteries, cool blood in the veins is rewarmed before it returns to the heart, thus helping the deer maintain a warm core temperature.

Feeling the chill

Red deer often suffer from a shortage of food in winter. Scientists have discovered that in order to reduce their energy requirements, red deer may allow their peripheral body temperature to drop significantly during periods of inactivity—most notably in the early hours of the morning. This drop in temperature is accompanied by a significant decrease in heart rate. These bouts of metabolic slow-down are similar to the torpor seen in hibernating mammals such as bears and dormice. The phenomenon long went unnoticed because it lasts just a few hours at a time.

Digestive and excretory systems

Deer feed by grazing (cropping grass and other plants growing on the ground) and browsing (plucking foliage from trees and shrubs). In both cases they use their muscular tongue to help grasp vegetation, which is then nipped off with the incisor teeth. The teeth may also be used to scrape or gouge the bark from trees to get to the softer, more easily digested tissue beneath. This tissue, known as cambium, contains the plant's water and nutrient transport systems—thus it can be a useful source of moisture and sugars in winter when green leaves of fresh grass are scarce.

Deer are ruminants. They specialize in eating vegetable matter, which can be difficult to digest. Like other ruminants, such as cattle and sheep, deer have a large, complex stomach, with four different chambers. Food is chewed, stored in the stomach, then regurgitated and rechewed. The true stomach or abomasum is adapted for digesting the sorts of materials found inside plant cells. Before food reaches the abomasum it has to be processed in such a way that these cell contents are released—this is where the other three stomach chambers (the rumen, reticulum, and omasum) come in.

Microorganisms and digestion

Plant cell walls are made of cellulose—a complex sugar that cannot be broken down by basic mammalian digestion. Deer and other ruminants therefore recruit colonies of

COMPARE the digestive system of the red deer with that of a non-ruminant herbivore such as a **ZEBRA** or a carnivore such as a **LION**. Herbivores have longer intestines than carnivores because plant material is more difficult to digest.

CONNECTIONS

▼ **Red deer**

The red deer's ruminant digestive system has a stomach with four chambers. This complex system is necessary to obtain the maximum amount of nutrient from the deer's plant-based diet.

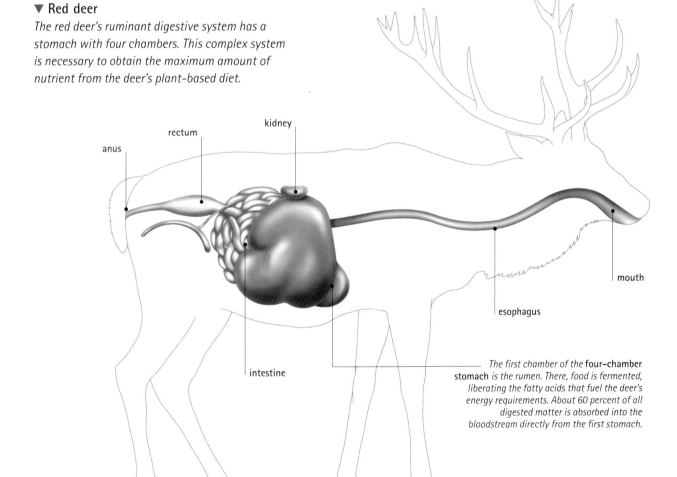

kidney

rectum

anus

mouth

esophagus

intestine

The first chamber of the **four-chamber stomach** *is the rumen. There, food is fermented, liberating the fatty acids that fuel the deer's energy requirements. About 60 percent of all digested matter is absorbed into the bloodstream directly from the first stomach.*

Four stomachs

Each of the four compartments in the deer's stomach has a slightly different character. The first chamber, the rumen, is large, occupying a large proportion of the deer's left side. Food from the rumen can be regurgitated into the mouth and rechewed. The lining of the next chamber, the reticulum, is covered with a honeycomb of ridges. The third chamber, the omasum, contains many large folds with horny bumps called papillae. The walls of the omasum, including the folds, are highly muscular, and food inside is squeezed and churned thoroughly before passing to the true stomach or abomasum, where the real process of digestion begins and the nutrients released from plant cells are attacked by acidic juices secreted from the stomach lining.

microorganisms to help do the job for them. These microorganisms include a variety of bacteria and single-cell protozoans, which reproduce continually in the first three chambers of the stomach. Not only do these microorganisms help release useful nutrients from the deer's food; they also ultimately become food themselves—millions of them pass into the true stomach with every meal, providing a valuable source of protein.

From the stomach, partially digested material passes into the small intestine, where secretions from the liver and pancreas help speed up the digestive process. True deer do not have a gallbladder. Musk deer, which are classified in the separate family Moschidae, do have a gallbladder.

Deer dung
Deer feces are expelled in clusters of pellets. Typically, these have a fine texture, with few signs of recognizable plant material, demonstrating the efficiency of ruminant digestion. The dung of nonruminant large herbivores, such as horses, contains large quantities of undigested plant matter.

◀ *Like other deer, white-tailed deer have long legs and a long neck. They can balance briefly on their hind legs and use their long neck to reach up for fruit on the lower branches of trees.*

IN FOCUS

Dietary supplements

The digestive system of deer is unquestionably adapted for processing vegetable material. However, deer will occasionally eat meat. As with many other herbivores, deer faced with poor grazing will take advantage of carrion—chewing on the carcasses of dead animals and sometimes devouring them completely. Animal tissue contains large amounts of protein and minerals that are not easily obtained from a restricted vegetarian diet. Similarly, stags also often chew their own cast-off antlers in order to regain some of the nutrients invested in them.

Reproductive system

Several species of deer, including muntjac and brocket deer, have been known to breed year-round. However, most species, including the red deer, are strongly seasonal and focus their breeding effort on an intensive period of competition and courtship known as the rut.

The breeding season in red deer lasts two to three months in the fall. Male red deer spend the summer feeding intensively and interacting peacefully with one another. However, as the rut approaches, levels of the hormone testosterone begin to rise and a number of physiological and behavioral changes take place. The neck muscles enlarge, and the hair on the neck grows long and shaggy to enhance the effect. The antlers are stripped of velvet and are honed and polished by being thrashed against vegetation or scraped on the ground. The males begin posturing and displaying to one another, showing off their strength.

At the same time, females gather in traditional locations, where they observe the

◀ A red deer calf suckles every two or three hours in the first few days of its life, but the frequency reduces as it gets older. The calf is weaned off its mother's milk after five to seven months. When the calf is very young, the mother eats the newborn's feces to prevent the scent attracting predators.

Caribou—females with antlers

Caribou are the only species in which both males and females have antlers. For a species that has to cope with some of the harshest conditions endured by any deer to invest so heavily in extravagant headgear suggests that there must be a significant advantage in having antlers. Female reindeer do not fight for mates, but they may have to fight for food. In winter food can be hard to find, and a reindeer may have to spend a lot of time and effort scraping away snow and ice to expose a patch of grass or lichen. One deer cannot afford to lose this to another deer, and the large antlers will act as a deterrent. The fact that female reindeer keep their antlers throughout the winter, long after males have shed theirs, supports this theory.

antics of the males. Rival males strut and bellow and wallow in mud and urine. Smaller, weaker individuals rarely get a chance to mate. Closely matched large males spend a long time sizing each other up, and if neither backs down they will resort to a test of brute strength, lowering their heads and locking antlers. From this position, each male attempts to wrestle the other to the ground. In serious fights, antlers may be broken, but they will regrow the following year.

Having proved his dominance, a male wins the right to a harem of females, but he must continue to defend them from the attentions of other males for the rest of the season. After two months of constant stress, such a male may be exhausted and half starved. Typically, he may also have suffered 30 or 40 antler punctures from fights with other stags. These wounds must be allowed to heal before the stag returns to the herd and social life.

Biologists working with mule deer discovered that some males avoided rutting, or fighting other deer, for several years. The scientists called these deer "shirkers." Since they saved used less energy and avoided costly wounding, the "shirkers" eventually became "super bucks"—with a very large body and very large antlers. When they eventually started rutting with other males these features gave them a big advantage over other males and they were able to mate with females readily.

Number of young

Female deer have two pairs of mammary glands and can thus theoretically nurse as many as four young simultaneously. However, females of most species rear only one young at a time. Exceptions to this rule are roe deer and white-tailed deer, which frequently raise twins;

▶ LOCKING ANTLERS
When two stags fight, they risk breaking their antlers. Fortunately, antlers regrow each year.

and Chinese water deer, which can give birth to as many as five or even six young, though it is rare for this many to survive to adulthood.

Development

Young deer are born fully furred and are able to stand almost immediately, but they are unable to walk far, so the mother leaves them lying concealed in long grass or other vegetation while she moves away to feed. For protection, the youngster relies on its dappled

▲ *Fallow deer mate in the fall. The doe usually gives birth to a single fawn between late May and mid-June.*

coat and its ability to remain perfectly still. Newborn caribou are able to follow their mother very soon after birth.

Weaning brings about important changes in the structure of the young deer's digestive system. At birth, the first three chambers of the stomach are very small—only the true stomach, or abomasum, is properly developed. This is all the youngster needs to process its simple diet of milk. The fore stomachs (rumen, reticulum, and omasum) begin to develop only when the juvenile deer begins to try solid foods.

AMY-JANE BEER

FURTHER READING AND RESEARCH

Geist, V. 1998. *Deer of the World: Their Evolution, Behavior, and Ecology.* Stackpole Books: Mechanicsburg, PA.

Nowak, R. 1999. *Walker's Mammals of the World,* 6th ed. Johns Hopkins University Press: Baltimore, MD.

Rue, L. L. 2004. *The Encyclopedia of Deer.* Voyageur Press: Stillwater, MN.

North American Game Species www.bowhunting.net/NAspecies/elk2.html

IN FOCUS

Boys and girls

Male red deer are called stags; females are called hinds; and the young are called calves. In other species, however, different names may apply. For example, male, female, and young fallow deer are bucks, does, and fawns respectively. Moose are called bulls, cows, and calves. *Hart* is an old word for a mature male deer—usually a red deer more than five years old.

Reproductive system

One of the essential characteristics of life is the ability to reproduce. In fact, all organisms that are living today exist because their parents and the parents before them, in generations going back to the first living being, were able to reproduce. The ones that failed to reproduce left no offspring and have no descendants.

DNA and RNA

All the information that is needed to reproduce a complex organism is contained within its DNA (deoxyribonucleic acid). Sections of DNA, called genes, contain code to build individual proteins. This code is copied into a closely related molecule called messenger RNA (ribonucleic acid; mRNA). The mRNA is then "read" by ribosomes, organelles within cells. Ribosomes are "machines" that translate the code in the mRNA to the corresponding sequence of different amino acids—the building blocks of proteins. Proteins, in turn, are functional units that carry out many of the tasks an organism needs to survive, grow, and reproduce.

However, DNA may not have been present at the dawn of life, although RNA may have been. The discovery that RNA molecules can help make other RNA molecules has led to the theory that the first forms of life on Earth may have been formed from self-replicating strands of RNA. By producing copies of themselves, these self-replicating RNA molecules would have become ever more abundant. Accidental errors introduced during the replication resulted in the occurrence of a variety of different RNA molecules, some of which started to carry the necessary information to produce proteins. This "knowledge" made the replication of RNA faster and more efficient. Because of the improved efficiency, RNA molecules that carried this information became much more numerous than the molecules that did not.

With time, an ever-increasing number of proteins were encoded by the RNA molecule, and at each step those RNA molecules that were

the most successful in reproducing themselves became the most abundant. As the raw materials to build these systems started to become limited, the less successful forms disappeared. Thus, very early in the history of life on Earth, reproduction may have become a vehicle for evolution; every time a change was introduced to the RNA that enabled it to be replicated more efficiently, this change tended to prevail in the lineages of future generations. At some point in time, RNA started to be copied to DNA, which took over as the storage medium for genetic information. This DNA encoded for proteins that speeded up its reproduction, and it is the ancestor of all life today.

Life-forms that are more effective in reproducing become more numerous, and those that are less successful fade. This process has created an amazing diversity of strategies to reproduce effectively, including many seemingly strange and fascinating features.

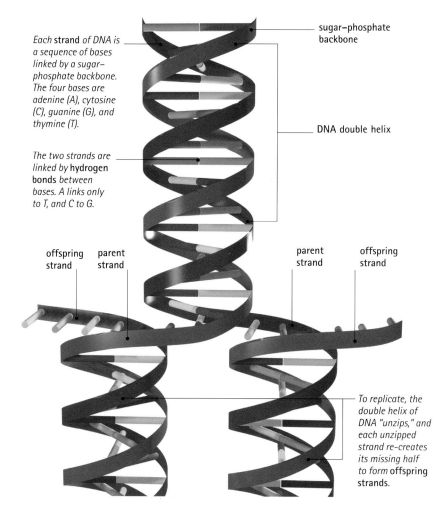

Each strand of DNA is a sequence of bases linked by a sugar-phosphate backbone. The four bases are adenine (A), cytosine (C), guanine (G), and thymine (T).

The two strands are linked by hydrogen bonds between bases. A links only to T, and C to G.

sugar–phosphate backbone

DNA double helix

offspring strand parent strand

parent strand offspring strand

To replicate, the double helix of DNA "unzips," and each unzipped strand re-creates its missing half to form offspring strands.

▶ DEOXYRIBONUCLEIC ACID (DNA)

All life, except some RNA viruses, uses DNA as the storage medium for genetic information. DNA is a long, threadlike molecule, composed of two intertwined strands. Sections of DNA, called genes, contain the code to build individual proteins.

If the driving force behind reproduction is to spread the genetic material of an individual, then the most efficient form of reproduction could be thought to be one where an exact copy, or a clone, of the individual is produced. This form of reproduction, which takes place without mating and in which the offspring is identical to its single parent, is called asexual reproduction. Many species reproduce asexually, and asexual reproduction has several advantages over sexual reproduction, which involves the combination of genetic material from two individuals into the offspring. First, asexual reproduction is the most direct and efficient way to propagate the genes (units of inheritance) of an individual. Second, asexual reproduction requires no mating to produce offspring, saving resources that otherwise would be put into courtship, for example. The third advantage is that there is no need to find a mate. This advantage is probably the reason why asexual reproduction is common among sessile animals (those that cannot move about) and those that live in dispersed populations where the chances of finding a mate are small.

Sexual reproduction

Despite the advantages of asexual reproduction, most multicellular organisms are able to reproduce sexually. During sexual reproduction, sex cells, or gametes, called eggs and sperm are produced by females and males, respectively. Almost all the body cells of most sexually reproducing organisms (exceptions include some plants) are diploid. Diploid cells contain two complete copies of the genome, which is the complete genetic information contained within DNA in a cell's nucleus. One copy of the genome comes from the mother and one from the father.

In contrast, eggs and sperm are haploid, with only a single copy of the genome. In sexual reproduction, two haploid gametes from two individuals fuse to make a diploid zygote, which then grows into an embryo. Sexual reproduction is universal among complex organisms, and this fact suggests that it brings an evolutionary advantage. By mixing genetic material from two parents, sexual

▲ Reproduction is the production of individuals similar to the parent or parents. In sexual reproduction, which occurs in most plants and animals, including cheetahs (above), male and female sex cells fuse to make offspring with genes from both parents.

reproduction produces genetic variation among offspring. The environment is unpredictable and liable to change drastically, causing some individuals to die. Others are likely to survive, because their particular combination of genes makes them able to cope with change. The chance that these offspring will be able to reproduce and pass on their genes is probably far greater than if all the offspring were asexual clones of one parent. To repeat: the fact that sexual reproduction is common clearly indicates that it must bring a huge evolutionary advantage.

FEATURED SYSTEMS

ASEXUAL REPRODUCTION This type of reproduction involves the production of offspring with only one parent. Asexual reproduction may be a part of a complex life cycle that also involves sexual reproduction at some stage. Alternatively, asexual reproduction in a species may continue for many generations or be the only type of reproduction. *See pages 986–989.*

SEXUAL REPRODUCTION During sexual reproduction two sex cells, or gametes, from two individuals fuse to start the development of a new individual that contains genetic material from both parents. *See pages 990–993.*

REPRODUCTIVE STRUCTURES IN ANIMALS Sexual reproduction requires structures and behaviors that enable the sex cells from two individuals to meet. The variety of structures and behaviors in animals and plants is incredibly large. *See pages 994–999.*

REPRODUCTIVE STRUCTURES IN PLANTS Asexual and sexual reproduction in plants requires specialized structures. *See pages 1000–1003.*

DEVELOPMENT OF THE ZYGOTE After fertilization, the zygote develops into an embryo, which continues developing, often within the mother. *See pages 1004–1005.*

Asexual reproduction

CONNECTIONS

COMPARE asexual reproduction in a **JELLYFISH** with that in a *SEA ANEMONE*.

COMPARE the regeneration of lost body parts in a *STARFISH* and a *CRAB*.

Asexual reproduction is widespread. Many organisms have complex life cycles, which may involve sexual reproduction at one stage and asexual reproduction at another. Different ways of reproducing asexually include fission, regeneration, and parthenogenesis.

Fission is the division of an organism into two organisms. In binary fission, a single-cell organism, such as a diatom, undergoes cell division to form two almost equal "offspring" cells. More complex forms of fission occur in multicellular organisms. Fission can take place at almost any stage in the life cycle of an organism. Adults can divide by fission through budding or stolonization (the growth of side stems to form new plants). In polyembryony, embryos form asexually in a plant seed or when an animal zygote splits, for example as in identical twins. Larval replication is fission during a larval life stage. It is relatively common among parasitic invertebrates with a complex life cycle.

Many jellyfish and some hydrozoans (such as obelias) alternate between life stages with sexual and asexual reproduction. The sexually reproducing form is the jellyfish stage, also called a medusa. The asexually reproducing larval form of jellyfish is called a polyp, the body shape that is normally associated with hydrozoans, such as *Hydra*. The medusa is the dominant phase in the life cycle of jellyfish, whereas hydrozoans spend most of their life in the polyp form. In both groups, however, the medusa exists as either a female or a male, which mates to form free-swimming larvae, or planulae. The larvae settle on the seabed, where they transform into the polyp form. In jellyfish, the polyp is called a scyphistoma. This repeatedly divides itself transversely, to form a structure called a strobila, which is composed of a stack of saucerlike structures called ephyrae. These break off from the strobila, one at a time, to form young medusae. All the medusae formed from a single

▶ **Whiptail lizards**
These lizards live in the arid grasslands of the southern United States and northern Mexico. In some populations of this species, there are only females, and young are produced parthenogenically—from unfertilized eggs. To activate egg development, the lizards pair up, undergo a rough courtship, and perform a sexual act in which they place their cloacas (reproductive openings) together. However, since neither animal is male, no sperm are transferred from one to the other.

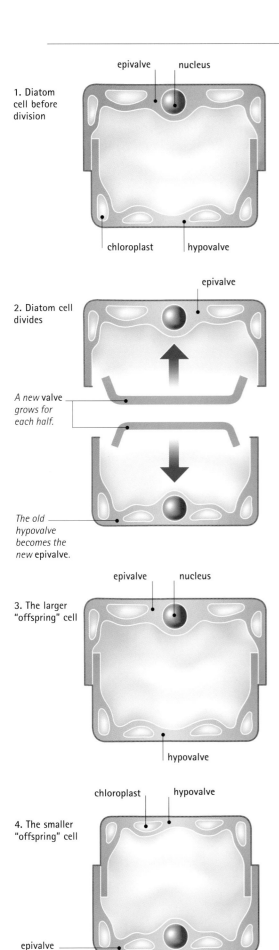

1. Diatom cell before division

epivalve   nucleus

chloroplast   hypovalve

2. Diatom cell divides

epivalve

A new valve grows for each half.

The old hypovalve becomes the new epivalve.

3. The larger "offspring" cell

epivalve   nucleus

hypovalve

4. The smaller "offspring" cell

chloroplast   hypovalve

epivalve

Asexual reproduction of blood flukes

Blood flukes are parasitic flatworms that cause an illness called schistosomiasis, or bilharziasis, in humans living in some tropical regions. The adult blood flukes live in veins of the intestine or bladder, depending on the species of fluke, where they mate and the females lay eggs. These eggs leave the host with the feces or urine. If the eggs end up in water, they hatch as free-swimming larvae miracidia. If a miracidium penetrates a freshwater snail, it becomes a sporocyst. The sporocyst reproduces asexually by fission to produce numerous tadpolelike larvae, called cercariae, which leave the snail. The cercaria are free-swimming, and if they meet human skin they penetrate it. After entering a human, they are carried in the bloodstream to the lungs, the liver, and finally the veins of the intestine or bladder, where they mature sexually and begin a new life cycle.

◀ **Diatom**
A diatom is a single-cell organism that reproduces both asexually and sexually. To reproduce asexually, the cell divides in two to form two "offspring cells." The cells are unequal in size, owing to the nature of the diatom's complex, silica-rich cell wall.

strobila are genetically identical and the same sex, because they are products of asexual reproduction. However, the medusae live independently and can reproduce sexually.

Some hydrozoans produce free-swimming medusae, but the freshwater hydra does not have this stage. Instead, the freshwater hydra reaches adulthood and reproduces sexually in the form of a polyp. However, the freshwater hydra also reproduces asexually. In the hydra, a new individual forms as an outgrowth from the body stalk of an adult individual. Through

▶ **Moon jellyfish**
Asexual reproduction occurs in the polyp stage of the jellyfish life cycle. The polyp, or scyphistoma, divides asexually to form a strobila—a stack of saucerlike structures called ephyrae. One by one, ephyrae break off to become medusae— free-swimming forms of the jellyfish.

Polychaete worms: Only the tails have sex

In the Syllidae family of polychaete worms, there are species in which parts of the body break off and engage in sexual reproduction. In the palolo worm, for example, a segment in the middle of the body grows into a second head, and the animal divides into two parts. The front part of the worm regenerates its mid and tail region and continues to live. The newly generated head is in charge of the rear end of the "parent" worm, which swells with sperm or eggs. The rear sections of many palolo worms swim to the surface, where they search for a mate and—if successful—spawn and die. Spawning is risky because the aggregation of worms attracts predators. However, since the front part of the worm remains hidden in sediments on the seabed, the only thing it risks is its tail.

this budding process, a complete hydra is generated. This asexually produced clone eventually breaks off from the parent animal.

Regeneration is the regrowth of body parts from a lost piece of the body. In some cases, this process can result in the generation of two or even several new organisms. If a starfish loses one of its arms, a new one can grow back. Moreover, the amputated arm—unless it was eaten or otherwise completely destroyed—can grow into a complete starfish.

Parthenogenesis, which means "virgin origin," is the process by which an egg develops into an intact individual without first being fertilized by sperm. Parthenogenesis is relatively common throughout the animal kingdom, and it is the only form of asexual reproduction that is

▶ *In some dandelions, seeds are produced asexually. In this type of reproduction, called apomixis, pollination and fertilization do not occur. Seeds develop from ovules containing diploid cells, rather than from the fusion of haploid egg and pollen cells.*

IN FOCUS

Alternation of generations in water fleas

Many species alternate between parthenogenesis (asexual reproduction from unfertilized eggs) and sexual reproduction. For example, in water fleas, parthenogenic females are produced when living conditions are favorable. When conditions worsen, males are produced parthenogenically. These males mate with females, which then produce eggs contained within a thick-walled capsule, or ephippium. When the female loses its exoskeleton, the ephippium is released and sinks to the bed of the lake in which the water fleas live. Ephippia can survive freezing and drying and can lie dormant for more than 100 years. They can even pass unscathed through the gut of a predator. When living conditions improve, the eggs hatch, giving rise to young water fleas.

eggs inside female water flea

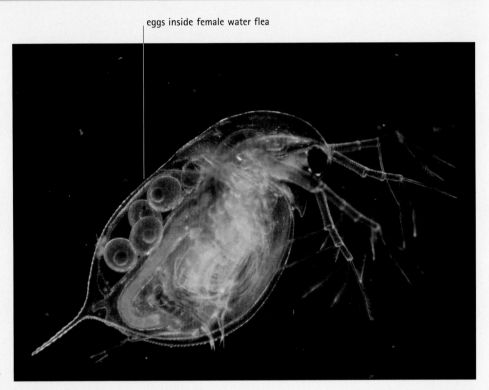

▲ *When conditions are favorable, water fleas produce female young by parthenogenesis (a form of asexual reproduction). If conditions worsen, males are produced. The males then mate with females, which produce eggs that lie dormant until conditions improve again.*

known to occur in vertebrates. Parthenogenic reproduction sometimes requires sexual activity even if this does not lead to fertilization of the egg. For example, egg development in whiptail lizards, which live in the southern United States, is triggered by courtship. However, some populations of whiptail lizards are made up entirely of parthenogenic females. In order to activate the development of eggs, some females (which have already laid eggs) behave as males: they court the female-acting partner and even engage in mating. As both partners are females, no sperm are exchanged, and the development of the eggs is triggered by the act of mating alone.

Asexual reproduction in plants

In plants, asexual reproduction is common. It is often divided into two types: apomictic and vegetative. Apomictic reproduction involves

offspring produced by unfertilized flowers and is comparable to parthenogenesis in animals. In a similar way, vegetative reproduction is comparable to fission in animals and involves the generation of new plants from parts of the plant that are not the germ cells. These include "runners" such as stolons, which are side stems that run from the parent plant above the soil and lay down roots from which new plants arise. Strawberries are a well-known example of a plant that produces stolons. In many plants the runners emanate from the parent plant in the soil, in which case they are referred to as rhizomes. A potato is the enlarged nutrient-filled tip of a rhizome, and humans have known for centuries that planting a potato, which is a stem and not a root, gives rise to new potato plants. In many plants, almost any part of the plant, including the stem and leaves, can generate a new complete individual plant.

Sexual reproduction

CONNECTIONS

COMPARE sex determination in a *HUMAN* with that in a *HONEYBEE*. In humans, it is a result of sex chromosomes: females have two X chromosomes, and males have an X and a Y chromosome. In honeybees, sex is determined by the number of sets of chromosomes: females have two sets (diploid) and males have just one set (haploid).

Sexual reproduction is based on the presence of females and males within a species. Females produce relatively large and immobile gametes (reproductive cells) called eggs, or ova; males produce sperm, or spermatozoons, which are smaller, mobile, and usually much more numerous.

The DNA in each cell of a eukaryotic organism (all life-forms except bacteria and viruses) is efficiently packed into dense structures called chromosomes. Chromosomes can be divided into autosomes and sex chromosomes. The latter differ between the sexes and contain the genes that lead to sex differentiation. In mammals and many other animals, the female and male sex chromosomes are called X and Y chromosomes, respectively.

Cells can have either a single copy of each chromosome or two almost identical copies. If a cell has only one copy of each chromosome, it is said to be haploid (n). During sexual

Sex determination in various animals		
	male	female
birds	ZZ	ZW
grasshoppers	XO	XX
humans, fruit flies	XY	XX
honeybees	haploid	diploid

reproduction, two gametes—an egg and a sperm cell—fuse to form a zygote, which has a double set of chromosomes, one coming from the egg and the other from the sperm. Cells that have a double set of chromosomes are called diploid (2n).

In many animals and plants, the major stage in the organism's life cycle is made up almost entirely of diploid cells, with only the gametes being haploid. A diploid zygote is created when two gametes fuse. This is called fertilization, and it usually occurs sometime

▶ Sexual reproduction in animals involves the fusion of sex cells—a sperm and an egg, or ovum—to produce new individuals. Sperm are much smaller than eggs and are mobile, swimming to the large immobile egg. This false-color micrograph shows a human sperm penetrating an egg. Magnified 700 times.

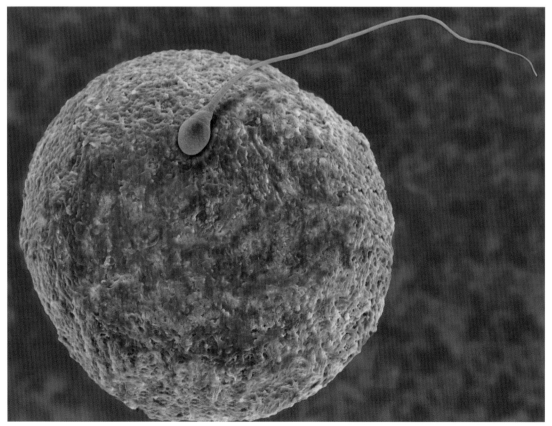

after mating, or copulation. Haploid gametes are formed from diploid cells, through a special kind of cell division called meiosis. Usually when cells divide, the chromosomes replicate and each offspring cell obtains the same chromosome setup as the parent cell. In this way, a diploid parent cell will produce two diploid offspring cells, and a haploid parent cell produces two haploid offspring cells; this type of cell division is called mitosis. In organisms in which the adults are diploid, haploid eggs or sperm are formed through meiosis. Meiosis actually involves two cellular divisions, although the DNA is copied only once. In this process the chromosome number is reduced from diploid to haploid.

Sex determination

In mammals, sex is genetically determined. A setup of two X chromosomes gives rise to a female, whereas the combination of an X and a Y chromosome results in a male. However, not all species follow this pattern. For example, in birds and some fish the presence of two sex chromosomes of the same kind (the ZZ setup) produces a male, and two different kinds of sex chromosomes (the ZW setup) leads to the development of a female.

The sex of an animal can also be determined by factors other than sex chromosomes. One example is environmental sex determination (ESD). In alligators and some other reptiles, the sex of an offspring is determined by the temperature at which the egg develops.

The vast majority of cells in an animal or a plant are differentiated for various specialized functions, such as building a skeleton, or taking up nutrients from the gut or in a root. These cells are called somatic cells. In contrast, the function of germinal cells is to propagate their DNA by the formation of new individuals. In animals, the germinal cells are defined very early in the embryonic development of an individual. They migrate from tissue close to the rear end of the gut, where they first appear, to the location of the developing ovaries (in females) or testes (in males).

In vertebrates, the female and male gonads (ovaries and testes) and sex organs develop from the same structures of the embryo. Early in the development of a vertebrate, there is no visual difference between females and males.

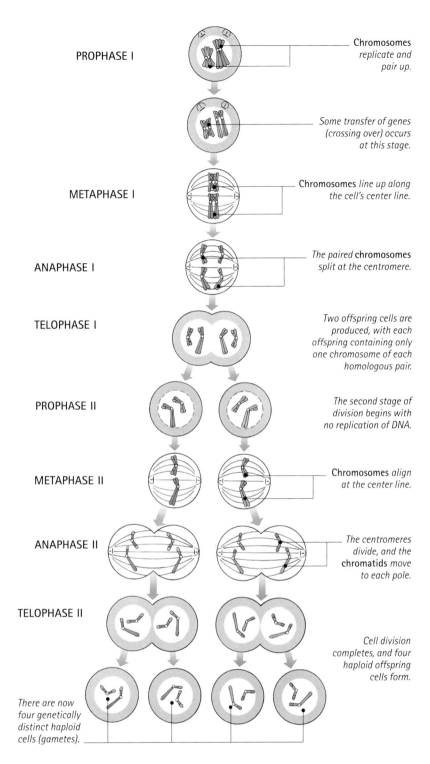

PROPHASE I — Chromosomes *replicate and pair up.*

Some transfer of genes (crossing over) occurs at this stage.

METAPHASE I — Chromosomes *line up along the cell's center line.*

ANAPHASE I — The paired **chromosomes** *split at the centromere.*

TELOPHASE I — *Two offspring cells are produced, with each offspring containing only one chromosome of each homologous pair.*

PROPHASE II — *The second stage of division begins with no replication of DNA.*

METAPHASE II — Chromosomes *align at the center line.*

ANAPHASE II — *The centromeres divide, and the* **chromatids** *move to each pole.*

TELOPHASE II — *Cell division completes, and four haploid offspring cells form.*

There are now four genetically distinct haploid cells (gametes).

The development into testes is dependent on a gene on the Y chromosome. Absence of the protein encoded by this gene results in the formation of a female reproductive system.

It is not only the gonads that differ between females and males; there are also differences between the brains of males and females. The development of "brain gender" is important because the brain directs behavior and controls

▲ MEIOSIS

Sexual reproduction involves the production of four haploid gametes (sex cells) from one diploid germ cell, in a two-phase process called meiosis.

GENETICS

Male-inducing genes

In humans, the testes-determining factor (TDF) is encoded by the SRY gene, which is located on the Y chromosome. If TDF is present, the undifferentiated embryonic gonads will develop into testes, and if TDF is absent, the gonads will become ovaries. When the testes are formed they start to produce the masculinizing hormone, testosterone, which stimulates the development of male sex organs. Testosterone also depresses the formation of female body parts, including the cells that otherwise would develop into breasts. The SRY gene is central to male development, but there are several other genes that are also believed to contribute to the differentiation between females and males.

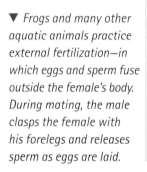

▼ *Frogs and many other aquatic animals practice external fertilization—in which eggs and sperm fuse outside the female's body. During mating, the male clasps the female with his forelegs and releases sperm as eggs are laid.*

functions involved in reproduction, and these behaviors and functions are sometimes very different in males and females. Sex-determining genes probably influence the gender of the brain directly and also indirectly by promoting the secretion of sex hormones.

Egg and sperm development

When the germinal cells arrive at the gonads early in the development of the embryo, they first divide by mitosis. In females, these cells become primary oocytes. The primary oocytes enter the first cell division of meiosis but then stop their development, remaining dormant until they are released from the ovaries during ovulation in the adult female. In humans, the egg does not complete the second cell division of meiosis until it is fertilized by a sperm.

In males, the number of germinal cells, called spermatogonia, in the fetal testes continues to increase by mitosis until birth. The division of diploid spermatogonia then stops until it is

Alternation of generations in plants

In animals, the haploid generation is short-lived and consists of eggs and sperm. In plants and fungi, the haploid generation may last much longer. Multicellular diploid individuals from these groups of organisms may give rise to multicellular haploid individuals, and the haploid generation is sometimes the dominant phase of the life cycle, as in mosses. In plants, the haploid generation is called the gametophyte. The gametophyte may have female and male organs. Male and female gametes (sex cells) produced by the gametophyte's reproductive organs fuse to give rise to the sporophyte, or diploid generation. The sporophyte then produces haploid gametes through meiosis. The haploid gametes germinate to grow into a gametophyte plant.

▶ **FERN LIFE CYCLE**
The life cycle of ferns alternates between a diploid sporophyte, which produces haploid spores; and a haploid gametophyte, which produces haploid sex cells by mitosis.

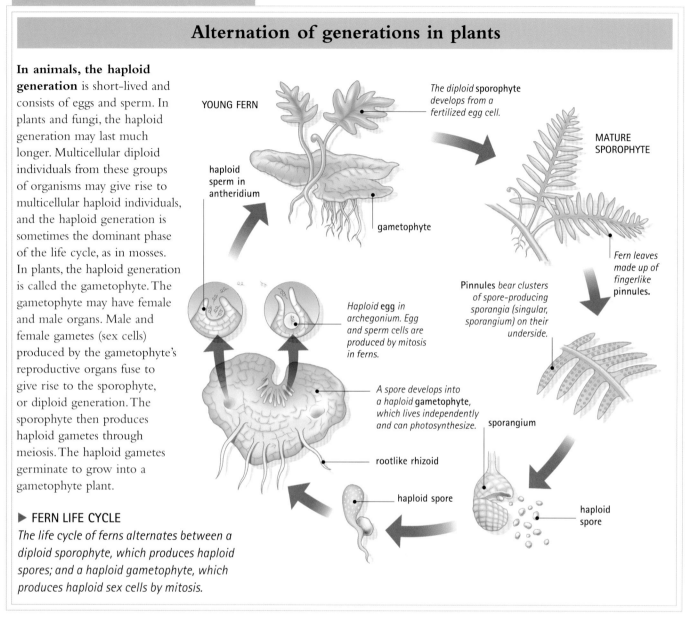

YOUNG FERN

haploid sperm in antheridium

gametophyte

The diploid sporophyte develops from a fertilized egg cell.

MATURE SPOROPHYTE

Fern leaves made up of fingerlike pinnules.

Pinnules bear clusters of spore-producing sporangia (singular, sporangium) on their underside.

Haploid egg in archegonium. Egg and sperm cells are produced by mitosis in ferns.

A spore develops into a haploid gametophyte, which lives independently and can photosynthesize.

sporangium

rootlike rhizoid

haploid spore

haploid spore

resumed at puberty when the germ cells, now called primary spermatocytes, start to undergo meiosis to produce a constant supply of cells called spermatids, which develop into sperm. Mature sperm are very small and contain, in addition to the haploid genetic material from the male, only what is absolutely required to deliver this material to the egg.

A generalized sperm has a small head, which contains the genetic material. In addition, there is a pouch, or acrosome gland, on the tip of the sperm head, and this is filled with a protein that helps the sperm penetrate the egg. The sperm head is followed by the sperm's midsection, or

body, which is packed with mitochondria. These provide the energy required to swim using the sperm's long tail—or in some species several tails. In many water-living species, eggs and sperm are released into the water, and the sperm fertilize the eggs outside the female's body; this is called external fertilization. This type of fertilization is difficult in land-living species partially because the sperm move by swimming. Therefore, in land-living animals fertilization typically occurs inside the female's body where body fluids can provide the medium for sperm and egg to meet. This type of fertilization is called internal fertilization.

Reproductive structures in animals

CONNECTIONS

COMPARE the claspers of a male *HAMMERHEAD SHARK* with the penis of a *DOLPHIN* and the spermatophore of *LOBSTER*.

Compare the brood pouch of a male *SEA HORSE* with the uterus of a female *COELACANTH*.

The sex organs, or testes, of most male mammals (except elephants, bats, and aquatic mammals) are located in a skin pouch, or scrotum outside the body cavity. Sperm probably develop best at a temperature slightly lower than that of the body core. Muscles in the scrotum are able to lift or lower the testes to vary their distance from the body and therefore maintain a constant temperature. Production of sperm (spermatogenesis) takes place in long, highly coiled tubes, called seminiferous tubules, which are located in the center of each testis. The seminiferous tubules have thick walls in which the sperm cells develop. In these walls, there are also Sertoli cells, which provide support and nutrients for the developing sperm cells. Sertoli cells also produce chemical substances, including hormones, that control the formation of sperm and sexual development in the body as a whole. The most immature diploid germ cells (spermatogonia) are located along the edges of the seminiferous tubules, and the more differentiated spermatids are nearer to the hollow center of the tubule with their tails pointing toward the center.

Between the seminiferous tubules there is another cell type, which produces hormones (chemicals that influence the function of cells

▶ **MALE REPRODUCTIVE SYSTEM AND FORMATION OF SPERM**
Human
Spermatogenesis, or sperm production, occurs in the seminiferous tubules of the testes. Diploid germinal cells divide by meiosis to produce haploid, tailed sperm. Sertoli cells assist spermatogenesis by providing support and nutrients. Sperm are ejaculated through the urethra during sexual intercourse.

▶ **SPERMATAZOON**
Human
Mitrochondria in the sperm's tail produce energy for the lashing flagellum, which propels the sperm. The acrosome at the front of the head allows the sperm to penetrate an egg.

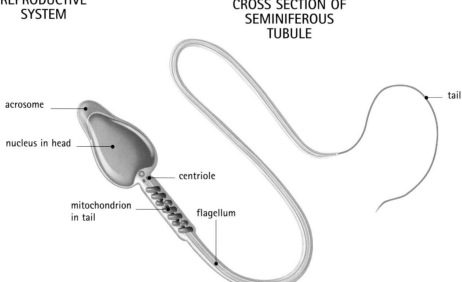

MALE REPRODUCTIVE SYSTEM: bladder, vas deferens, epididymus, seminal vesicle, erectile tissue, prostate gland, urethra, epididymus, testis, glans penis

TESTIS: seminiferous tubules

CROSS SECTION OF SEMINIFEROUS TUBULE

DETAIL OF TUBULE: Sertoli cell, germinal cell, diploid spermatocyte, haploid spermatocyte, spermatid, spermatozoon

SPERMATAZOON: acrosome, nucleus in head, centriole, mitochondrion in tail, flagellum, tail

elsewhere in the body). These cells are called Leydig cells, and they make the masculinizing hormone, testosterone, which stimulates the Sertoli cells inside the seminiferous tubules to direct the development of sperm.

When the sperm cells are almost mature they are released into the central lumen of the seminiferous tubules. They are moved along the seminiferous tubules to a storage area, or epididymis, which is located next to the testis. There, the sperm cells go through their final maturation and are now ready to be mixed with other components in a fluid called semen.

Sperm and semen are expelled from the body via tubes called the vas deferens and the urethra, the latter ending at the tip of the penis. Along the way, other fluid secretions are added by the bulbourethral glands, the seminal vesicles, and the prostate gland. To allow sperm to enter the female's reproductive tract, the male's penis is inserted into the vagina of the female.

Female mammalian reproductive system

The ovaries of a female mammal are located in the posterior (rear) part of the body cavity. In the ovaries, the developing egg is surrounded by supportive cells. Some of these cells, called theca cells and granulosa cells, produce and secrete feminizing sex hormones. Collectively, the egg and its supporting cells make up a follicle.

At ovulation, eggs are released from the ovaries into the funnel-like opening of the oviduct (the fallopian tubule). Hairlike cilia lining the inner walls of the oviduct sweep the eggs toward the uterus. If sperm are present in the female's reproductive tract, they will usually intercept the egg in the upper part of the oviduct. To get there the sperm will have traveled from the upper parts of the vagina, where they were

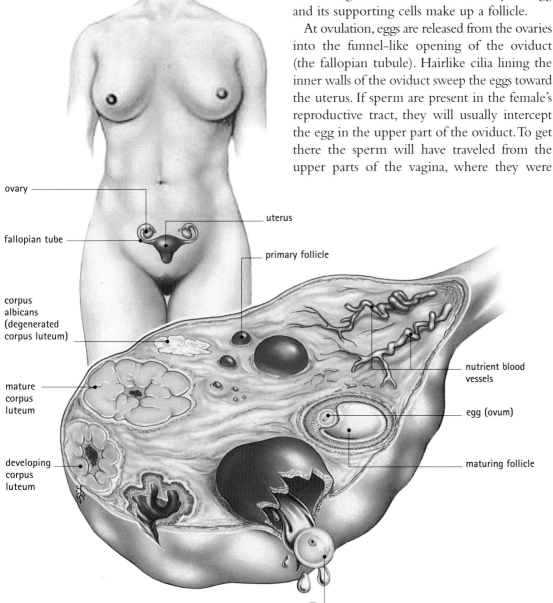

ovary

fallopian tube

uterus

primary follicle

corpus albicans (degenerated corpus luteum)

mature corpus luteum

developing corpus luteum

nutrient blood vessels

egg (ovum)

maturing follicle

mature egg (ovulation)

◀ EGG DEVELOPMENT IN HUMANS
Human females have two ovaries, each containing follicles, in which eggs develop. When an egg is released (ovulation), the empty follicle develops into a corpus luteum. This secretes the homone progesterone, which causes the wall of the uterus to thicken in preparation for implantation of a fertilized egg. If this does not occur, the corpus luteum degenerates.

995

ejaculated from the male, through the cervix, which separates the vagina from the uterus, and along the uterus into the oviduct.

In humans, the opening of the vagina (and the opening of the urethra through which urine passes) is framed by two pairs of skin folds: the labia minor and labia major. The labia minor are thin and richly supplied with sensory nerves that respond strongly to touch. Where the two labia minor meet in front of the urethra is a knob of erectile tissue, called the clitoris, which has the same origin in the embryo as the penis in males. Around the labia minor are the labia major,

which have thicker skin and protect the delicate tissues situated underneath. In human infants the opening to the vagina is partly covered by a thin membrane called the hymen. The hymen often ruptures or stretches in childhood, for example during physical exercise; or this may happen during the first sexual intercourse.

Vertebrate sex hormones

Hormones are chemicals that are produced in one part of the body and are carried in the bloodstream to other parts of the body where they have an effect. In vertebrates, the release

▼ **Female emperor dragonfly**
Eggs develop in the two ovaries and pass along the oviduct to the bursa. Dragonflies practice internal fertilization. The male has a structure called an aedeagus, which is used to introduce sperm into the female's bursa. In females, sperm is stored in spermathecae and released to fertilize eggs.

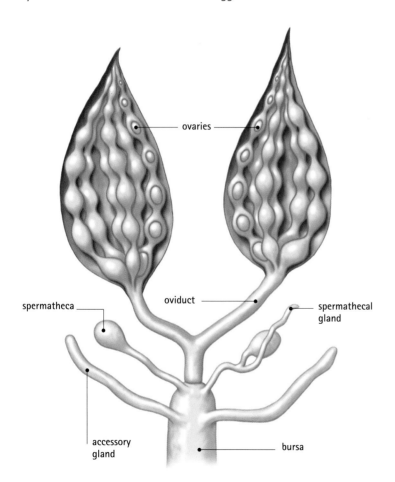

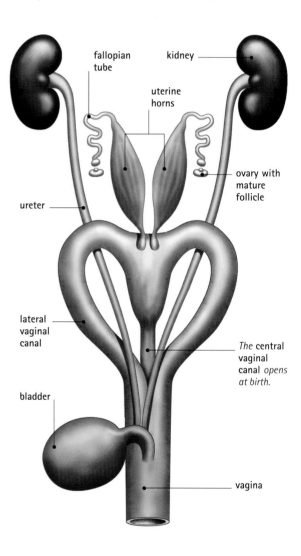

▲ **Female kangaroo**
Like many other female mammals, kangaroos have two ovaries in which eggs develop. However, kangaroos are unique in having two wombs, or uterine horns, and two curved vaginas. When a kangaroo is ready to give birth, a new opening develops, through which the joey passes.

of hormones that direct sexual reproduction is coordinated by an area of the brain called the hypothalamus. The hypothalamus receives information from other parts of the brain and produces signals that decide the timing of sexual maturation and ovulation, or the release of eggs from the ovaries.

The hypothalamus releases a hormone called gonadotropic-releasing hormone (GnRH). This hormone acts on a small gland, the pituitary gland, or hypophysis, which lies just below the hypothalamus. The pituitary gland produces several hormones, and two of these, follicle-

Timing of reproduction

In many animals, sexual reproduction occurs only during a limited period of time every year, and it is therefore important that females and males are prepared to reproduce at the same time. Day length and temperature are common cues for sexual maturation in animals that live in nontropical regions. In the tropics, there may be other triggers, such as the phase of the moon, drought, rain, or flood. Chemical signals (pheromones) released by individuals that are ready to breed are also very important in coordinating sexual reproduction within a population.

▼ Male hammerhead shark
Sharks have a single opening, called a cloaca, through which they excrete wastes and reproduce. Sperm produced in the testes travels along Wolffian ducts to the cloaca. The male transfers sperm into the female using one of his claspers, which has a groove along which the sperm travel.

▼ Male Jackson's chameleon
All lizards and snakes have two penises, called hemipenises, one on either side of the cloaca. During mating, sperm pass down the spermatic ducts to one of the hemipenises, which is inserted into the female's cloaca. Fertilization of eggs by sperm takes place in the female's oviducts.

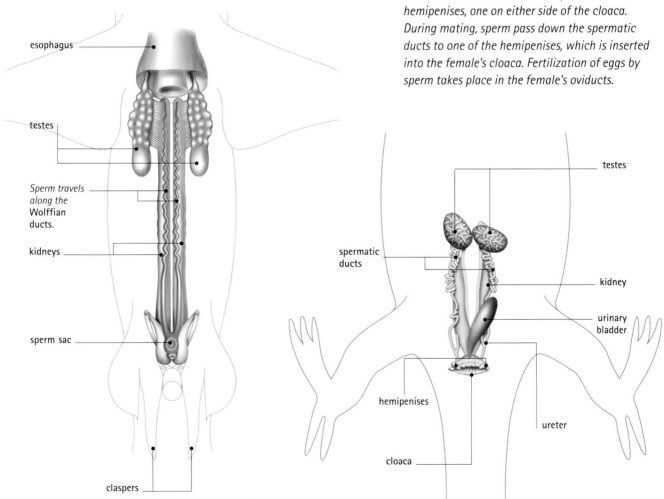

esophagus

testes

Sperm travels along the Wolffian ducts.

kidneys

sperm sac

claspers

testes

spermatic ducts

kidney

urinary bladder

hemipenises

ureter

cloaca

997

Hermaphroditism

Hermaphroditism is the state of an individual having both male and female sex organs. There are two kinds of hermaphroditism: simultaneous, in which the organism has male and female organs at the same time; and sequential, in which the animal changes from male to female, or vice versa. Some coral reef fish are hermaphrodites. Clown fish, for example, live closely associated with sea anemones. Only the biggest clown fish associated with a suitable sea anemone is female. She breeds with the second

biggest fish, which is the only adult male. If the female disappears, the sexually mature male will change sex and become female; one of the previously immature fish will mature and become the new breeding male.

In some fish species a single breeding male has a harem of females, and each individual starts its adult life as a female and may later change sex to become a male. The queen angelfish follows this strategy. It would make little sense to be a small male queen angelfish because the dominant male does not tolerate

other males in his territory. Therefore, most of the other fish in the group are females until the dominant male disappears. When this occurs, the largest of the females changes sex and becomes a male.

▼ Clown fish can change sex. If the largest fish in a group (the breeding female) dies, the only breeding male then changes sex to become the new female. One of the immature fish becomes the new male.

stimulating hormone (FSH) and luteinizing hormone (LH), respond to GnRH. In males, LH stimulates the production of testosterone in the Leydig cells of the testes. Testosterone stimulates the growth of the testes and promotes the bodily changes associated with male puberty. These changes vary greatly among different species and include an increase in muscle mass

and hair growth, changes in coloration, and, in humans, a deepening of the voice. FSH acts with testosterone on the Sertoli cells inside the seminiferous tubules to stimulate the production and maturation of sperm cells.

In female mammals, FSH and LH stimulate production in the ovaries of several female sex hormones, collectively called estrogens, which

Testosterone and birdsong

Every breeding season, male birds sing to claim their territory, compete with other males for dominance, and attract females. As the breeding season approaches, and before the males start to sing, the testes begin to expand. The growing testes produce more and more testosterone. The testosterone stimulates further testis development and sperm maturation and produces changes in the brain. For example, testosterone stimulates the growth of several parts of the brain required for singing.

▶ *Lincoln's sparrow is widespread in North America. The male bird has a rich, gurgling song that it uses to attract a mate.*

in turn stimulate the maturation of the eggs and the development of female sex characteristics throughout the body.

In mammals, these female sex characteristics include the growth of the mammary glands (breasts) and, in humans, widening of the hips and distribution of fat tissue to the buttocks and thighs. In vertebrates that lay eggs or otherwise have embryos that are not nourished from the mother through a placenta, estrogens are also responsible for stimulating the production of a yolk-protein precursor in the liver. This protein is then transported in the bloodstream to the ovaries, where it is incorporated into the developing egg and converted to yolk.

Another key function of estrogen hormones in female mammals is to stimulate the growth of the inside lining of the uterus. This lining is called the endometrium. It is within the endometrium that the embryo embeds and develops into a fetus.

Corpus luteum and progesterone

As the egg or eggs (depending on the species) mature and the follicle grows, the production of estrogens increases. At a certain point in time (12 days after the start of menstruation in

humans), the pituitary gland produces a surge of the hormones LH and FSH. This surge leads to the ovulation of a mature egg or eggs from the ovaries. The ruptured follicle is left behind in the ovary, and it transforms into a structure called a corpus luteum. The corpus luteum continues to produce estrogen and a hormone called progesterone. The endometrium depends on estrogens and progesterone for continued growth and maintenance, in preparation for the implantation of a fertilized egg.

Shedding of the endometrium

If the egg or eggs do not become fertilized, the corpus luteum withers away (this happens within two weeks after ovulation in humans). As a result of the drop in levels of estrogen and progesterone that follows, the endometrium starts to break down.

In most species of mammals, the endometrium is absorbed by the body, but in certain primates, such as humans, chimpanzees, and gorillas, the thickened lining of the uterus is shed through the vagina, a process called menstruation. If an egg is successfully fertilized by sperm, growth and development of the embryo continue in the endometrium.

Reproductive structures in plants

COMPARE
pollination in
ORCHIDS with that
in **SAGUARO** and
MARSH GRASS.
Orchids are mainly
pollinated by
insects, such as
bees, and moths;
the saguaro cactus
is pollinated by
bats and also bees;
and marsh grass is
wind-pollinated.

Plant reproduction differs fundamentally from animal reproduction because plants have a multicellular haploid phase of the life cycle as well as a multicellular diploid phase. In some groups of plants, such as mosses and liverworts (bryophytes), the haploid stage, which is called the gametophyte because it bears gamete-producing sex organs, dominates the life cycle; and in other groups (notably flowering plants, or angiosperms) the diploid stage, which is called the sporophyte because it produces spores, dominates.

Gymnosperms were the first seed-bearing plants to evolve. There are perhaps fewer than 750 gymnosperm species alive, but some of these are very widespread. The most familiar gymnosperms are conifer trees, which include pines and spruces. The sporophytic scale tissue of the seed cone and the gametophyte inside constitute the ovule. The term *gymnosperm*, meaning "naked seed," refers to the ovule of the plants in this group, which is not protected by an ovary or fruit tissue as it is in flowering plants, or angiosperms. *Angiosperm* means "enclosed seed."

Flowering plants

Angiosperms are the most diverse group of living plants, with more than 257,000 species. Angiosperms have flowers for at least part of the year, and these are the plants' sexual organs. A typical flower consists of sepals, petals, male stamens, and a female carpel. The carpel contains the ovules and seeds. Stamens typically consist of a thin stalk, or filament, with an anther at the end. The anther contains

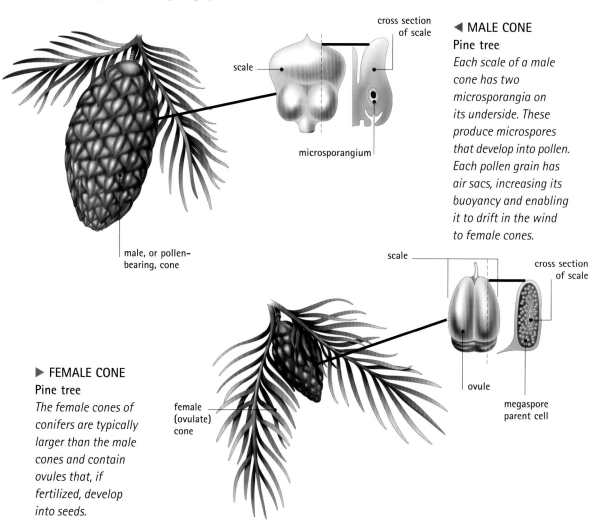

cross section
of scale

scale

microsporangium

◀ MALE CONE
Pine tree
*Each scale of a male
cone has two
microsporangia on
its underside. These
produce microspores
that develop into pollen.
Each pollen grain has
air sacs, increasing its
buoyancy and enabling
it to drift in the wind
to female cones.*

male, or pollen-
bearing, cone

scale

cross section
of scale

ovule

megaspore
parent cell

▶ FEMALE CONE
Pine tree
*The female cones of
conifers are typically
larger than the male
cones and contain
ovules that, if
fertilized, develop
into seeds.*

female
(ovulate)
cone

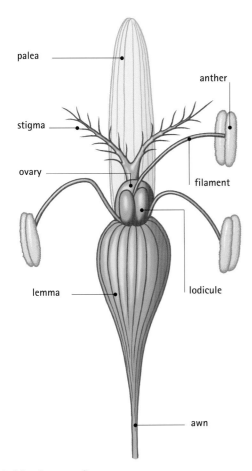

palea

anther

stigma

ovary

filament

lemma

lodicule

awn

▲ **Marsh grass flower**
Grass flowers are wind-pollinated and do not need showy petals to attract insects. Two scales, the palea and lemma, initially enclose the male and female organs. The scales are split apart by the swelling of two small bodies called lodicules.

IN FOCUS

Gamete production in flowering plants

In the anthers of a flower, diploid cells called microsporocytes divide meiotically to form haploid microspores, which in turn develop into pollen grains. Pollen grains have two nuclei: a generative nucleus and a tube nucleus. (After pollination, the generative nucleus divides mitotically to make two sperm nuclei.)

Within the ovule of the carpel, a diploid megaspore parent cell divides through meiosis to produce four haploid megaspores. Each megaspore divides mitotically four times to produce eight haploid nuclei, all enclosed within one single cell, called a megagametocyte. Two groups of three nuclei each migrate to opposite ends of the megagametocyte and two nuclei stay in the middle. In this rearrangement of nuclei, cell walls are built around the six nuclei that are located at the ends of the megagametocyte. One of these cells becomes the egg (which, if pollinated, fuses with a sperm nucleus from the pollen). The two nuclei in the center of the megametocyte become enclosed within the same cell wall to produce a single cell containing two polar nuclei (which, if pollinated, fuses with the other sperm nucleus from the pollen).

the pollen-producing microsporangia, or pollen sacs. Megasporangia are housed in the carpel. In a single flower, there may be one or more carpels, which may be fused. The carpel has a stigma where pollen may attach. The stigma is often at the end of a style, which connects the stigma to the ovary at the base of the whole structure. The ovary, style, and stigma are together called the carpel.

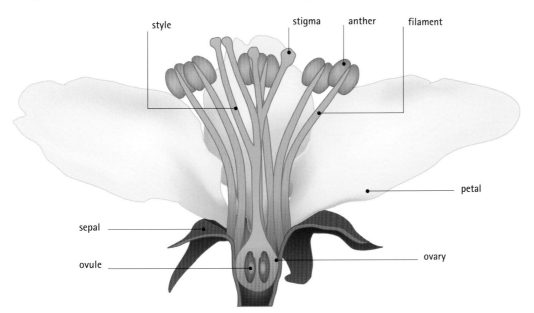

style

stigma   anther   filament

petal

sepal

ovule

ovary

◄ **Apple flower**
Flowers with both male and female reproductive organs, such as apple, are called "perfect." The female parts are called carpels and are situated in the center of the flower. Each carpel is made up of a stigma, style, and ovary. The male parts, which surround the carpels, are called stamens; each one is composed of an anther and filament.

▲ *Many flowers are pollinated by bats. Such flowers open, or remain open, during the night and emit strong, fruity smells to attract bats. Above, a long-tongued bat, which occurs in North and South America, visits a cup-and-saucer plant.*

IN FOCUS

Fertilization in flowering plants

When a pollen grain lands on the stigma of a compatible carpel, it develops a pollen tube and burrows into the stigma and down the style until it encounters the ovule. Chemical signals—probably released by cells in the ovule—serve as a guide for the pollen tube. These chemicals lead the pollen tube to a small opening in the ovule. At first, the tube contains two haploid nuclei. One is the generative nucleus and the other is the tube nucleus. The haploid generative nucleus divides by mitosis to produce two haploid sperm nuclei. Both sperm nuclei are released into the ovule; one of the sperm nuclei fuses with the egg to produce the zygote of the new diploid sporophyte generation. Curiously, the other sperm nucleus fuses with the central cell in the ovule that already contains two nuclei. The fertilized central cell now contains three sets of chromosomes; it is triploid (3n). The triploid cell rapidly divides through mitosis and produces a tissue that serves as nutrient for the developing plant embryo. The process is called double fertilization and is unique to angiosperms.

However, there is considerable variation in this generalized flower structure. For example, many angiosperms have female-only or male-only flowers, lacking carpels or stamens, respectively. Flowers that have both carpels and stamens are called "perfect," and those that lack either male or female structures are called "imperfect." If an individual plant has both female and male flowers, it is called dioecious. Otherwise, plants are either female or male and are called monoecious. In either case, it is the sporophyte generation that produces flowers. The gametophyte of flowering plants is highly reduced—the female gametophyte consists of only seven cells.

Unlike other plant groups, such as mosses and ferns, angiosperms and gymnosperms are not dependent on water for the male gamete to travel to the egg. Instead, the male gametes of angiosperms and gymnosperms sit in pollen grains that may be carried by wind or by animals (for example, insects, bats, and birds) that make contact with the flowers.

Following fertilization, there is a program of cell division and development in the embryo and in the tissues that surround it. The cell layers surrounding the ovule develop into a protective seed coat. The embryo itself develops into a stage in which buds develop that will form the seed leaves. Later, the seed will lose most of its water content (up to 95 percent), and it stops growing and becomes dormant. The embryo remains in this quiet state until conditions are favorable for it to germinate.

IN FOCUS

Reproduction in a gymnosperm

Pine trees provide a good example of gymnosperm reproduction. Pine trees have two types of cones: seed cones and smaller strobili. Seed cones produce megaspores, which are the female gametophytes; and the strobili produce microspores, which are male gametophytes. Microspores develop into pollen grains, which travel on the wind to sticky seed cones. The pollen grain then germinates to form a pollen tube, which makes its way—over several months—to the female gametophyte. One of the pollen cells then divides to produce two sperm cells, one of which fuses with the egg cell to form a zygote. Each scale of the seed cone may contain several eggs, which when fertilized can produce several zygotes. In most cases, only one of these survives. Each zygote develops into an embryo within a seed. Pine seeds are wing-shaped, so they can be carried by the wind away from the parent plant to grow in new location.

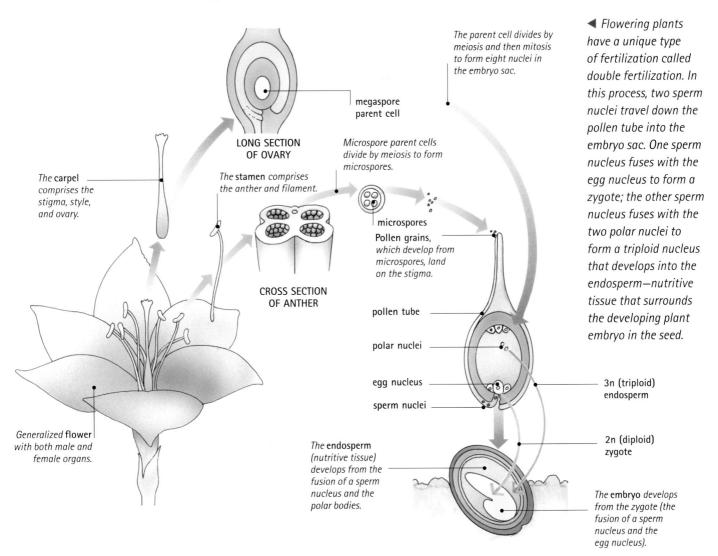

The parent cell divides by meiosis and then mitosis to form eight nuclei in the embryo sac.

megaspore parent cell

LONG SECTION OF OVARY

Microspore parent cells divide by meiosis to form microspores.

The carpel comprises the stigma, style, and ovary.

The stamen comprises the anther and filament.

microspores

Pollen grains, which develop from microspores, land on the stigma.

CROSS SECTION OF ANTHER

pollen tube

polar nuclei

egg nucleus

sperm nuclei

Generalized flower with both male and female organs.

The endosperm (nutritive tissue) develops from the fusion of a sperm nucleus and the polar bodies.

◀ Flowering plants have a unique type of fertilization called double fertilization. In this process, two sperm nuclei travel down the pollen tube into the embryo sac. One sperm nucleus fuses with the egg nucleus to form a zygote; the other sperm nucleus fuses with the two polar nuclei to form a triploid nucleus that develops into the endosperm—nutritive tissue that surrounds the developing plant embryo in the seed.

3n (triploid) endosperm

2n (diploid) zygote

The embryo develops from the zygote (the fusion of a sperm nucleus and the egg nucleus).

Development of the zygote

In animals, a sperm penetrates an egg using proteins stored in the acrosome to make a hole. The sperm then loses its tail, and the membrane that surrounds the egg cell nucleus breaks down, allowing the DNA of the two gametes to combine. The resulting zygote divides repeatedly without gaining size until a species-specific number of cells have been formed. This initial bout of cell divisions results in the formation of a ball of cells called a blastula. The blastula has a fluid-filled cavity, called the blastocoel, around which much of the embryo development occurs. The formation of a blastula marks the completion of the first major step in the development of an embryo.

The second milestone in the developmental process is gastrulation. During gastrulation, the wall of the embryo forms a pouch that reaches deep into the blastocoel. The opening to this pouch is the pore that in many animals, including all vertebrates, becomes the anus. The cavity inside the pouch will become the intestine, and the wall of the pouch (the endoderm) will also give rise to other organs, such as the liver and lungs. From this primitive gut, two pockets appear and pinch off, creating a hollow sac on either side of the gut. The two sacs are the coelomic vesicles, and the cells that make up the sacs (the mesoderm) will form

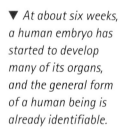

▼ *At about six weeks, a human embryo has started to develop many of its organs, and the general form of a human being is already identifiable.*

▲ *Female dogfish release egg sacs—commonly called mermaid's purses—in which embryos develop for several months before hatching. The embryos are nourished by a large reserve of yolk.*

internal tissues such as muscles, blood vessels, reproductive organs, and the inside lining of the body cavity. At this stage, the ball of cells is called a gastrula. The outer cell layer of the gastrula is called the ectoderm, and it will give rise to the body surface and superficial structures such as hair and nails, the nervous system, and some other tissues.

In vertebrates, the third major stage of zygote development is the formation of a nervous system. First, the notocord, which develops into the backbone, is laid down along the ceiling of the gastrula. The nervous system is formed from the ectoderm (outer cell layer) overlaying the future backbone. Above the rod of backbone cells, the ectoderm thickens and

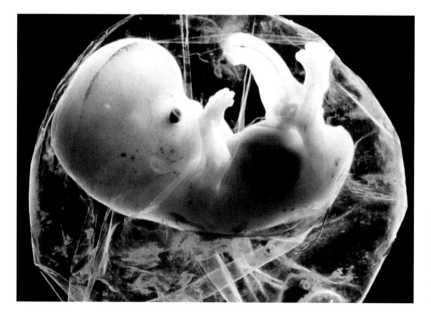

GENETICS

Hox genes

The differentiation of the body along the longitudinal axis is at least partially determined by a large family of genes, which in vertebrates are called Hox genes. The proteins that are encoded by these genes are present at different concentrations and in different combinations along the longitudinal body axis. This unequal distribution of Hox gene products decides where the brain, spinal cord, front limbs, hind limbs, and other body parts will form.

flattens to form a neural plate. Ridges develop along the longitudinal edges of the neural plate, and these ridges rise and fold toward the middle, where they fuse, generating a neural tube. The neural tube develops into the central nervous system. The front end of the neural tube becomes the brain, and the rest becomes the spinal cord.

During the formation of the neural tube, cells from the endoderm grow into longitudinally repeated segments on either side of the neural tube. These segments of endodermal tissue are called somites. Early in the development of the

▼ COMPARISON OF EMBRYOS
Amphibian, bird, and mammalian embryos divide, or cleave, at different rates and have structural differences right from the outset of development. For example, bird embryos quickly develop a large reserve of yolk.

embryo, the somites are more or less uniform, but they become successively more diverse. The somites later give rise to several of the structures that are unevenly distributed along the body, including the ribs, vertebrae, limbs, and muscle. Once the task of system and organ development has been accomplished, the remaining period of development is primarily characterized by growth. In humans, the final phase is called the fetal period.

Gestation and the placenta

In animals that bear live young, such as most mammals and some fish and reptiles, the period from fertilization of the egg until birth is called gestation. In mammals, the length of gestation is generally related the animal's life span. For example, in short-lived mammals, such as small rodents, gestation may last less than a month, whereas in long-lived species, such as elephants, it can be as long as 22 months. In humans, gestation is called pregnancy and takes about 40 weeks (9 months). The length of gestation is also related to the size of the fetal skull that will fit through the mother's pelvis during birth.

All mammals except monotremes (such as the platypus) and marsupials (such as kangaroos and the koala bear) are placental. The placenta is the organ by which the developing embryo attaches to the wall of the uterus. The placenta develops from the chorion (the outermost layer of cells of the blastula). It is attached to the lining of the uterus and connects to the growing fetus by the umbilical cord. Oxygen, nutrients, and antibodies pass across from the mother's blood—via a thin layer of cells—into the fetus's blood; and fetal waste products pass into the mother's blood for excretion by her kidneys and lungs. No direct mixing of fetal and maternal blood occurs. In this way, the fetus is nourished until birth. The placenta is then expelled in the afterbirth.

CHRISTER HOGSTRAND

	Amphibian	Bird	Mammal
AT FERTILIZATION	polar bodies	yolk mass	zona pellucida
FIRST CELL DIVISION		blastodisk (yolk-free region)	
BLASTULA (EXTERIOR)		yolk, blastoderm	
BLASTULA (VERTICAL SECTION)	blastocoel, subgeminal space	yolk	blastocyst cavity, inner cell mass

FURTHER READING AND RESEARCH

Marshall Graves, Jenny. 2004. *Sex, Genes, and Chromosomes.* Cambridge University Press: Cambridge, UK.

Purves, W. K., G. H. Orians, D. Sadava, and H. C. Heller. 2003. *Life: The Science of Biology,* 7th ed. W. H. Freeman and Company: New York.

Index